PREVENTION'S

Low-Fat, Low-Cost
Freezer Cookbook

Quick Dishes For and From the Freezer

by Sharon Sanders

PREVENTION
Health Books ™

Rodale Press, Inc.
Emmaus, Pennsylvania

OUR PURPOSE

"We inspire and enable people to improve
their lives and the world around them."

Prevention's Low-Fat, Low-Cost Freezer Cookbook
Editorial Staff

Managing Editor: Anne Egan
Editor: Sharon Sanders
Writing and Recipe Development: Mary Carroll
Senior Copy Editor: Kathy D. Everleth
Copy Editor: Marilyn Hauptly
Interior Designer: Tad Ware & Company, Inc.
Design Coordinator: Kristen Morgan Downey
Layout Designer: Pat Mast
Illustrator: Carolyn Vibbert
Photographer: Kurt Wilson
Food Stylist: Roscoe Betsill
Prop Stylist: Evelyne Barthelemy
Nutritional Consultant: Christin Loudon, R.D.
Manufacturing Coordinator: Melinda B. Rizzo
Office Manager: Roberta Mulliner
Office Staff: Julie Kehs, Bernadette Sauerwine

Rodale Health and Fitness Books

Vice-President and Editorial Director: Debora T. Yost
Executive Editor: Neil Wertheimer
Design and Production Director: Michael Ward
Research Manager: Ann Gossy Yermish
Copy Manager: Lisa D. Andruscavage
Production Manager: Robert V. Anderson Jr.
Studio Manager: Leslie M. Keefe
Book Manufacturing Director: Helen Clogston

In all Rodale Press cookbooks, our mission is to provide delicious and nutritious low-fat recipes. Our recipes also meet the standards of the Rodale Test Kitchen for dependability, ease, practicality, and, most of all, great taste. To give us your comments, call 1-800-848-4735.

For the best interactive guide to cooking, nutrition, weight loss, and fitness, visit our Web site at http://www.healthyideas.com

CONTENTS

INTRODUCTION

How would you like to save hundreds of dollars on food expenses each year? Well, the cash box is right in your kitchen. And this book gives you the key.

Nearly every household in America has a freezer attached to the refrigerator. And more than one-third of U.S. homes has a free-standing freezer. This underutilized home appliance can be a secret source of cold cash. In addition to real dollars, a freezer yields invaluable benefits for the smart meal manager.

First, you can eat more nutritiously with far less effort. Frozen single-ingredient foods, such as fruits and vegetables, frequently retain more nutrients than fresh counterparts that lose nutrients rapidly every day after harvest. Frozen fruits and vegetables are processed straight from the field or tree, locking in the vitamins and minerals so essential in warding off cancer, heart disease, and other illnesses.

Second, the freezer lets you spend less time on cooking and more time with family and friends or pursuing leisure activities. A two-pronged strategy doubles your effectiveness.

•Those recipes designated as "For the Freezer" are marked with this symbol. They're dishes that you can prepare in bulk to freeze for effortless no-cook dinners at a later date.

•Dishes earmarked as "From the Freezer" recipes are accompanied by this symbol. They're made of freezer and pantry staples and take only minutes to pull together.

Third, freezing is the food preservation method that least changes the color, texture, and flavor of foods. Unlike dried, pickled, or canned foods, frozen foods are often indistinguishable from their cooked unfrozen counterparts. When foods look and taste this appealing, meals are bound to be a hit.

Each recipe includes the price per serving and a nutritional analysis so you can monitor your monetary budget and your fat budget at a glance. Scores of tips on bulk buying, freezer packaging, and thawing foods guarantee that your freezer cuisine will taste as good as—or better than—fresh.

Sharon Sanders

CHAPTER 1

PUT A FREEZE ON RISING FOOD COSTS

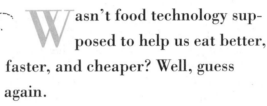

Wasn't food technology supposed to help us eat better,
faster, and cheaper? Well, guess
again.

Better? Fast-food carryout and
supermarket convenience meals are
no nutritional gift. Laden with fat,
sugar, and salt, they offer the most of what we should be
eating the least. Even so-called healthy prepared meals
are too often lacking in dietary fiber and essential vitamins and minerals.

Faster? How long does it take to wind through that in-
terminable line at the local burger joint? Or wait with
increasing irritation because your home is the last stop
on the pizza delivery person's route?

Cheaper? Highly processed foods always cost far
more than the sum of their ingredients. And when you
tally in the hidden costs of medical bills resulting from a
poor diet, the costs of commercial convenience become
too high to bear.

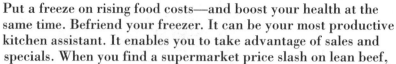

Put a freeze on rising food costs—and boost your health at the same time. Befriend your freezer. It can be your most productive kitchen assistant. It enables you to take advantage of sales and specials. When you find a supermarket price slash on lean beef, you can buy it in bulk, confident that the freezer will keep it until you're ready to use it. The same goes for that bushel of apples that you pick up for a song at the end-of-season farmers' market. With a freezer, it's easy to stock double batches of favorite meals for effortless heat-and-serve dinners.

Imagine the luxury of inviting friends over for a spur-of-the-moment dinner. Family dinners will also become more fun. With the main dish ready, there's time to create a special appetizer, salad, or dessert—or just relax after a hard day at work.

Take a new look at your freezer. It's been sitting in your kitchen for years like money in a noninterest-bearing account. It's time to make it earn its keep by yielding tasty dividends.

In this book, we share scores of time-, labor-, and money-saving tips and more than 220 recipes that are either "From the Freezer" or "For the Freezer." Those designated as "From" dishes—made primarily from frozen staples that you have on hand—are like getting fast cash from an automated teller machine. "For" dishes are like money in the bank—easy, satisfying favorites that pay dividends of effortless meals on nights when you don't feel like cooking at all.

Save Cold Cash

A freezer can drastically reduce your grocery bills. Experts say that it's easy to use the freezer efficiently once you know the basics of food selection and preparation.

Your freezer is your number one kitchen tool for avoiding highly processed foods and costly convenience items. We show you how to really utilize your freezer to create simple and delicious meals made with a variety of minimally processed foods that enhance well-being and lower the cost of each dish. Grains, legumes, vegetables, and fruits—flavored by small amounts of lean meats, seafood, and low-fat cheeses—are foundations of tasty low-cost cooking. Your freezer maintains their natural goodness until you need them as building blocks for healthful meals.

We teach you how to use these foods in money-saving, great-tasting, time-tested classics like Sweet-and-Sour Pork, and Shepherd's Pie as well as new offerings like Salsa Chicken Dinner and Spicy Beef Wraps. With smart use of your freezer, even poultry, meat, and seafood entrées needn't cost more than $2 per serving. Most dishes cost only pennies.

Each recipe includes nutritional data and the cost per serving, so you can use your freezer to plan meals, adjust for special health concerns, and budget all at the same time.

Prices are based on the lowest seasonal, bulk, and sale prices in Minneapolis supermarkets and include homemade yogurt, stock, and other items most often on hand in the low-cost kitchen. Prices in your area may vary.

Use this book as a road map to thrifty, healthy cooking and learn how easy it is to put great taste on your table every night with the help of your freezer.

Freeze In Nutrients

Did you ever fly home from vacation in California, Florida, or Mexico? You probably didn't feel too fresh after traveling all that distance. Well, imagine the condition of vegetables that cover those same thousands of miles in the back of a truck. When you think of how far foods travel to get to your table, it's no wonder that many nutrition experts recommend frozen foods when you can't get fresh, locally grown ingredients.

"Fresh fruits and vegetables live and respire, using up their nutrients," says Robert L. Shewfelt, Ph.D., professor of food science and technology at the University of Georgia in Athens. "The longer you store them, the more they use. Freezing captures those nutrients and maintains them for a long period of time."

Nutritionists recommend that we eat three to five servings of vegetables and two to four servings of fruits every day. A wide variety of nutritionally rich fruits and vegetables is easy to find year-round in your supermarket freezer section, which makes it easy to expand your dietary choices. If you're looking for variety to boost your health and still want to keep a handle on costs, rely on the freezer, says Amy Dacyczyn, thrift expert and author of *The Tightwad Gazette* series.

Size Up Savings

Depending on the size of your household, your freezer may occupy anything from the space on top of the refrigerator to a 20-cubic-foot chest in the basement or in a heated garage. Freezers come in many shapes and sizes. Is yours big enough? Two cubic feet of freezer space per family member is a good yardstick for gauging if your freezer is adequate as a money-saving kitchen tool.

3

10 AT ALL TIMES

If you have only a standard refrigerator-freezer, it's crucial to choose the essential staples that will serve you best. Keep these top 10 food items on hand in the freezer.

1. Boneless chicken breasts or turkey breast cutlets
2. Bread crumbs
3. Chopped onions
4. Cookie dough or baked cookies
5. Defatted chicken stock
6. Fruits
7. Pie crusts
8. Soups
9. Spaghetti sauce
10. Vegetables

A stand-alone freezer is the key to buying food in bulk and really saving on grocery bills, Dacyczyn says. She encourages even renters to own a freezer and maybe arrange to lock it in the basement. "To buy in bulk most effectively," she says, "it pays to have a separate freezer. Used for garden surplus as well, the savings far outweigh the cost of the unit or the electricity. The largest available freezer for home use costs less than $500 and uses about $8 worth of electricity a month. It could easily pay for itself within a year."

To keep a freezer humming, it's important that frost not accumulate on the walls and shelves. "If your freezer isn't a frost-free model, you'll help the unit maintain a proper temperature and use less electricity if you unplug and defrost the freezer before the frost buildup gets to ¼" thickness," says Earl Proulx, *Yankee* magazine fix-it columnist and author of *Yankee Magazine's Make It Last!*

Frost-free freezers are designed to defrost a bit during their cycles, a feature that saves defrosting time but can play havoc with poorly wrapped foods. If you have a frost-free freezer, use freezer-quality wrapping (see "Polar Protection" on page 8), which blocks both moisture and air, to protect your edible investments.

Proper location also contributes to the freezer's health and well-being. Freezers last longer if placed in a cool, dry, and well-ventilated room where the temperature stays above 40°F, says Proulx. Fluctuating temperatures cause problems for freezer coils.

Keep your freezer's internal temperature at 0°F (most home freezers operate on a range from 0° to 8°F). A freezer thermometer on duty allows you to monitor internal temperature and tell what part of your freezer is coldest, usually the back of the bottom shelf.

Shop Smart by Stocking Up

How you shop for bargains and stock your freezer can drastically reduce your grocery bills. Research shows that the national average we spend for food each week hovers around $125 for a family of four, but with shopping savvy, you can cut this figure in half or more without giving up any flavor or satisfaction.

Jonni McCoy, author of *Miserly Moms: Living on One Income in a Two-Income Economy*, uses smart-shopping techniques to lower her family's food budget to $40 a week. Here's her advice.

- Avoid shopping when you're in a hurry, hungry, or tired, because these are the times for impulse buying. Foods that you really save money on, like potatoes, flour, and dried beans, are not what grab your attention when you're rushed.
- Strategize your shopping trips to take advantage of supermarket sales on items that you can stock in the freezer.

McCoy relies on a price book—a pocket-size spiral notebook in which she records the cheapest price for each staple she buys and the store that offers that price. She checks the unit price, which is posted on small markers tacked to the supermarket shelves, to determine if a sale item is really a bargain. Surprisingly, a smaller size can occasionally be a better value than a large one.

Dacyczyn also strongly advocates a price book. In her price book, each staple gets a full page. This simple system is an essential step to slashing grocery bills. Write down prices while you're shopping or later from the store receipts. It will save you plenty when visiting a farm stand, farmers' market, or warehouse store, where sale items are in unfamiliar large sizes or in bulk units. That bushel of tomatoes that you're planning on freezing as spaghetti sauce may actually be cheaper bought at your supermarket's sale price than at the local farm stand.

5

Dacyczyn advises stockpiling staples by bargain-hunting during supermarket sales and summertime visits to local farm stands. Buy as much as you can reasonably afford and your freezer can easily store. You can get even better bargains when you buy bigger—a side of beef, 30 pounds of whole chickens, or a bushel of onions, for instance. McCoy calls meat-packing companies to see if she can get any discounts on large quantities, then organizes other families into an informal buying group.

Remain flexible so you can avoid running to the supermarket to replenish an item that's not on sale. Check your freezer instead. Dacyczyn's family may use up all the fresh garden tomatoes by mid-fall, so she'll switch to her own home-frozen tomatoes rather than purchase fresh tomatoes out of season. Remember that you always pay more for out-of-season convenience, and the quality may not match that of seasonal produce.

What—And How—To Freeze

When your bargain purchases are commercially frozen foods, such as single-ingredient vegetables or fruits, they are already packaged for you in durable freezer containers. Make sure that you pick up these items last on your shopping circuit, pack them together in the same bag or box for the drive home, and place them in your freezer as soon as you walk in the door.

Freezing foods maintains color, flavor, and texture a lot longer than other preservation methods, such as canning or drying. Most frozen foods—especially out of season—beat fresh for superior quality of taste, and texture, says Shirley Corriher, author of *Cook-Wise*. But to get the best results from foods you freeze at home, it pays to know a little about what happens when food freezes then thaws.

"All fresh food contains water, and as the water freezes, ice crystals form between the cells," Corriher explains. This causes water inside the cell to flow out, replacing the water outside that has been frozen and keeping liquid balance on both sides of a cell. As the outside water freezes more, the inside of the cell becomes more concentrated, more soaked with natural salts that rupture the cell membranes. "This explains why thawed meats can taste dry," says Corriher, "and crisp vegetables can become soft and limp."

The faster a food is frozen, the more palatable it will be, Corriher advises. Food that freezes quickly forms small ice crystals, which means less cell damage and moisture loss.

To freeze fast, says Corriher, first cool the food as much as you can before it goes into the freezer. Also known as flash-freezing, this is particularly effective for many foods such as muffins, boneless chicken breasts, raw or cooked burgers, cupcakes, pancakes, waffles, and berries. Flash-frozen foods are easy to remove in individual portions or measure by cupfuls without thawing the whole container.

To do this, place the food items on a tray or baking sheet in a single layer and without the pieces touching. Freeze for one to three hours, depending on the density of the food, until solid. Then pack it into a freezer-quality plastic bag.

To retain good texture in produce, cook the vegetable or fruit slightly before freezing it. "If you partially break down the cell structure before freezing, the food won't change as much in the freezer," says chef Michael Roberts, author of *Fresh from the Freezer*, who blanches vegetables in boiling water for anywhere from 30 seconds to several minutes. This prevents cell explosion during thawing, he says. Drain the blanched vegetables quickly and rinse or submerge them in cold water. Drain again and pat dry.

Fully cooked dishes, such as stews, soups, and casseroles, also need to be thoroughly chilled before freezing. Partially fill the sink with cold water, then set the hot cooking pot in the sink to cool. Allow the water to come to within one inch of the top of the pot. Adding ice cubes to the sink will accelerate the cooling. The idea is to cool the dish to room temperature as quickly as possible. After the dish is cooled, chill in the refrigerator for several hours before wrapping or packing into containers for the freezer.

If you snag a bargain, you can also freeze certain foods that aren't usually commercially frozen, such as tortillas, breads, and even certain dairy products (see "Freezing Dairy Products and Eggs" on page 179). Baked goods keep best in the freezer, far better than in the refrigerator. Part-skim ricotta cheese and low-fat shredding cheeses like Cheddar, Swiss, and Monterey Jack all freeze well. So does buttermilk. Low-fat sour cream and yogurt can be frozen if used for cooking after they are thawed. But don't attempt to freeze any fat-free dairy products, warns Corriher. Their stabilizers—the ingredients that create a smooth texture in place of fat—can break down.

High-starch foods, such as whole raw potatoes, don't freeze well on their own, says Corriher. Raw potatoes can be mixed with a cream or cheese sauce, which will help stabilize them. Shredded cooked potatoes freeze acceptably.

Avoid freezing sauces thickened with flour; they separate into

Put a Freeze on Rising Food Costs

"a sponge and a puddle" on thawing, says Corriher. If you're making a sauce or gravy specifically for the freezer, thicken it with cornstarch or instant tapioca.

Don't forget your freezer for storing bulk grains such as oatmeal, whole-wheat flour, and rice, says Dacyczyn. They keep for years in the freezer. Whole grains, in particular, will be protected from going rancid.

Most desserts freeze well, says Elinor Klivans, author of *Bake and Freeze Chocolate Desserts*. "Freezing means that you can bake one for now, one for later—no extra time spent, and you get twice as much. I love knowing that I have dessert ready when I'm entertaining." Klivans usually has brownies, pie crusts, cake layers, cookies, unbaked fruit pies and other fruit desserts, and meringue shells on hand.

Polar Protection

Who would venture into 0°F weather without dressing for it? Not your pampered packages of frozen foods. Wrapping foods properly for the freezer is critical if you expect to see your money's worth from them months down the road.

"Freezer air, being so cold, has the natural tendency to suck moisture from foods," says Klivans, who recommends moisture-proof, air-proof plastic wrap and heavy foil. Both are easy to find in the supermarket. Look for the words "freezer quality" on the box. Wrap foods first in freezer-quality plastic wrap, then in freezer-quality foil, Klivans advises. The plastic wrap blocks moisture; the foil blocks cold air. When you keep out these two cold-hearted villains, you prevent the flavor loss that results from freezer burn.

Wax paper, lightweight foil, lightweight plastic wrap, cellophane, dairy cartons, produce bags, and butcher paper don't provide a strong enough barrier to keep moisture and air away from food for more than a few days, Klivans adds.

Dacyczyn uses freezer-quality plastic bags and expels all the air before sealing, either by pressing it out of the bag with her hands or immersing the bag in water.

Ice-cube trays are a freezer essential for Roberts. He portions purees, sauces, pestos, leftover tomato paste, and even stock in ice-cube trays. Buy inexpensive plastic trays at discount stores. Coat the trays with no-stick spray to prevent stronger flavors like pesto or spaghetti sauce from permeating the plastic. Freeze the cubes solid, then flick them into a plastic bag, expelling all the air before sealing. Label and date the bag. Store in the freezer for up to six months. Roberts says that the cubes can be thawed individually and added to dishes. Or just pop the frozen cubes straight into a soup or stew pot.

Meats purchased in bulk or family packs are best frozen in smaller amounts, says Corriher, so repackage them into meal-size portions. Remove the meat from the supermarket trays and wrap in freezer-quality foil or freezer-quality plastic wrap to prevent freezer burn.

Partially preparing your staples before freezing saves effort at cooking time. When you get a bargain on whole chickens for 29 cents a pound, skin, bone, and otherwise prepare the chicken so that it's recipe-ready. For example, if your family loves stir-fries, stock your freezer with containers of sliced uncooked chicken. If broiled chicken with salsa is a favorite, have packages of boned and skinned breasts on hand. Do the same with turkey breast.

If meals in your house are typically for two or if family members eat in shifts because of outside activities, package soups, stews, lasagna, spaghetti and meatballs, and other staple dinners in small containers—the food will thaw faster, and the small size is more convenient for on-the-run dinners. These single-serving containers also make economical microwaveable lunches.

Label It or Lose It

Oh where, oh where are those baby limas you froze last summer? Ironically, the better the packaging, the harder it is to identify the food without a label. If you ever want to find anything in your freezer, you have to label it, says Klivans.

Even freezer fanatics confess that labeling is the hardest habit to begin but the one that's most beneficial. A minute spent identifying foods—while you still can—is 5 or 10 minutes saved when you want to quickly retrieve something. Always mark with waterproof pens or china marking pencils. Write your labels before applying them to the package so that you don't inadvertently puncture the wrapping with your pen. Make sure that you write the name of the item and the quantity. But don't stop there.

Some chefs use colored stickers to sort and quickly identify types of frozen foods: green for vegetables, blue for fruits, red for meats—an especially helpful tip for harder-to-organize chest freezers.

Dating foods also becomes second nature with a freezer storage system. Adding the month and year means that you'll know when

QUICK THAWS

It's six o'clock, and you're in a panic because you forgot to pull dinner ingredients from the freezer this morning. Don't worry, be *zappy*!

Thawing in the microwave is fast, simple, clean, and less worrisome as far as spoilage is concerned, says chef Michael Roberts, author of *Fresh from the Freezer*. And you don't have to plan ahead as you would to thaw food in the refrigerator overnight. Just follow the directions in your microwave instruction booklet. But be sure to cook the food immediately after thawing, especially raw meat, poultry, or seafood.

- Most frozen vegetables and small shrimp, scallops, and individually flash-frozen fish fillets cook in seconds in hot liquid. You can add them directly from the freezer to soups, stews, and other saucy dishes. Just add a few minutes to the allotted cooking time.

- Cooked seafood can be thawed by dipping in boiling water. Use a slotted spoon to keep the fish intact and drain well after dipping.

- Need a cupful of peas, spinach, or corn kernels for a recipe? Just scoop out the individually flash-frozen vegetables, place in a colander, and run hot water over them for 30 seconds. Drain well and add to your dish.

- Frozen nuts and home-frozen chopped herbs can be used directly from the freezer in any recipe, cooked or not. Let the dish stand for a few minutes at room temperature for the nuts or herbs to release their flavors.

to use up that special buy before it spoils. It's not wise management to buy chicken on sale for 29 cents a pound and then forget to use it, Dacyczyn says.

Manage Frozen Assets

Do you have to use sleight of hand to grab what you want and close the freezer door before you get buried under an avalanche of frozen food? Do you play 20 questions with mystery packages? If so, some easy management techniques can guide you through that vast frozen wasteland of unidentified objects to a jackpot of cold cash savings.

- Make quick use of frozen pita bread, sliced breads, and bagels without thawing. Simply pop them in the oven, toaster, or toaster oven to reheat for speedy sandwiches. Frozen pizza shells can also be topped and baked without thawing.

- For burritos and other tortilla dishes, you can pull frozen tortillas out of the freezer at the start of preparation time, and they will thaw by the time you've made the filling. Or thaw frozen tortillas quickly in the microwave.

- Homemade cooked dishes—such as soups, stews, and casseroles—that have been frozen in single-serving containers can be thawed and reheated in the microwave in minutes.

- Package your own homemade stocks or cooked beans in 1-cup portions for the freezer. You can conveniently remove just the amount you need for quick thawing in the microwave.

- Keep in mind that planning is essential for thawing larger items such as a whole chicken or a roast. Thaw overnight in the refrigerator. Cool temperatures keep the food from spoiling as it thaws completely. Keep the frozen food in its freezer packaging during thawing.

- Thawed foods are especially perishable, so use them up fast. And *never* thaw meats, poultry, fish, or similar food at room temperature. You're just asking for an unwelcome visit from dangerous bacteria.

View the freezer as a storage system. Much like office file cabinets or kitchen cupboards, it serves best when it's organized. To earn its keep, the freezer has to work well for your lifestyle.

Chefs who depend on the freezer for cost control and convenience recommend three simple practices.

1. Organize freezer items by type of food and how often you use them.
2. Package foods properly to keep them in prime condition.
3. Label and date all freezer packages.

It takes just a few minutes to synchronize these three steps into your routine, but each saves you time when you're in a

How Long Can I Keep It?

Many foods can be frozen for up to a year. Commercially frozen foods have a slightly longer freezer shelf life than foods frozen in a home kitchen. Likewise, larger foods, such as a whole turkey, will keep longer than smaller foods, such as turkey breast cutlets. If properly wrapped and stored in a

Food	Maximum Freezer Life (months)
MEAT, POULTRY, AND SEAFOOD (UNCOOKED)	
Beef, ground	4
Beef, roasts	8–12
Beef, steaks	8
Beef, stew meat	4
Chicken, cut up	10
Chicken, whole	12
Fish, fatty	3
Fish, lean	6
Lamb, chops	6
Lamb, leg roasts	9
Pork, chops	3
Pork, tenderloin, shoulder, or roasts	6
Shrimp	3
Turkey, cut up	6
Turkey, tenderloin, cutlets, or drumsticks	6
Turkey, whole	12
BAKED GOODS	
Cakes	6
Cookies, baked	4
Cookie dough	3

kitchen crunch. Plus, you reap the full value from everything you freeze.

A freezer operates most efficiently when it's fully packed, but space must be left around the air vents so that the cold air can circulate properly. To organize items efficiently, you might use extra baskets, bins, shelves, and rack dividers to sort your freezer space by type of food.

Take a cue from the frozen-food case at the supermarket to or-

freestanding freezer, which maintains a constant temperature of 0°F or lower, even home-frozen foods will show little appreciable loss in flavor, color, or texture. Consult the following table for the maximum freezer time for various foods.

Food	Maximum Freezer Life (months)
BAKED GOODS (CONTINUED)	
Fruit pies, unbaked	6
Muffins	4
Quick breads	4
Yeast breads	3
Yeast dough	1
PRODUCE	
Berries	4
Dried beans, cooked	6
Fruit, cut up and small whole	4
Vegetables, cut up and small whole, blanched	6
DAIRY AND EGGS	
Butter, reduced-calorie	1
Cheese, grated (Parmesan, Romano)	24
Cheese, ricotta	1
Cheese, shredded (Cheddar, Mozzarella, Swiss)	4
Egg whites, raw	12
Sour cream, low-fat	4

ganize your home freezer most efficiently. Store like items together. Klivans separates her freezer into sections for different types of baked goods. One shelf is for breads, another is for the cookies her family loves. A wire basket holds ingredients, such as small bags of nuts and frozen fruits.

Store the oldest foods toward the front so that they're visible when you open the freezer. Eat last year's blueberries before you go pick this year's crop. This is a simple stock rotation system

FREEZER POWER

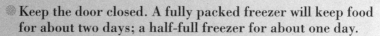

You've shopped smart, wrapped right, diligently labeled, and stocked your freezer well. But what do you do when that inevitable power outage happens?

❄ Keep the door closed. A fully packed freezer will keep food for about two days; a half-full freezer for about one day.

❄ If you know that the outage will last for more than one day, wrap the food in heavy blankets and store it in coolers. Use up what you can. One chef says, "Throw a huge party!" (Of course, if the power is out, you may have to do so at someone else's house.)

❄ For safety's sake, don't refreeze thawed raw meats, fish, or poultry. If you cook them—especially in a soup or stew—you may refreeze without loss of quality or fear of spoilage.

that's used in supermarkets to ensure that the first foods in are the first foods out.

Freeze only foods that your family enjoys. The spinach surprise that everyone rated "thumbs down" won't magically get a "thumbs up" after six months in the freezer, no matter how economical it seems to save it.

Dacyczyn and other freezer fanatics post an inventory sheet on the freezer door. It's a tool to plan menus, keep a balanced assortment of foods on hand, and use foods before they're past their peak.

Create an inventory sheet with a wipe-off board or a homemade chart, to add and delete items as needed. A glance will tell you immediately what's on deck for dinner. One chef even makes a sketch of the freezer compartments, photocopies a dozen, then posts one copy on the door. He pencils in the foods he has, adding and subtracting as needed, then replaces the chart with another photocopy when it gets too marked up.

Plan for Convenience

Your freezer can help you reduce not only prices but also time spent in the kitchen. With advance planning and stockpiling, you can even come home from work and put together a great meal from the freezer in less than 30 minutes. "Onions cooked in large batches then frozen in small amounts can cut 5 or 10 minutes off every recipe you cook that week," says Roberts. Tomato sauce, cooked grains and beans, even marinated chicken can find a home in your freezer and be only minutes away from your dinner table. Roberts recommends learning how to build many of your menus around a well-stocked freezer to make cooking faster and fun.

A small amount of planning each weekend keeps you in control of your health and wealth. Instead of panicking and calling for pizza delivery when things get hectic, you just calmly thaw and serve a satisfying homemade dinner. Many of the recipes in this book are superconvenient—made with frozen ingredients that don't even need to be thawed before cooking. It's easy to steer clear of the money trap of commercial convenience because you'll have economical, convenient, home-cooked meals at your fingertips.

Advance planning also lets you know when to replenish your freezer stock and which staples to look for on sale because you know you'll run out in two weeks. Shopping trips become simpler too, honed down to a short list and scheduled on the day of the week that offers the best selection, triple coupons, and sales. And you'll know when to visit farmers' markets, warehouse stores, or buying clubs for certain bulk buys.

A well-stocked freezer allows you to average the cost of your meals over the course of a week. For example, if you spend $1 a serving on broiled chicken breasts on Wednesday, you might serve cheese enchiladas at 50 cents a serving for Thursday's dinner. You can stock up on several months' worth of protein values. You can choose a variety of lean meats or explore a wide range of grain- or bean-based ethnic meals—often the best cuisines for the most economical eating.

When menu planning, set aside time to browse through the recipes in this book and choose several dinner entrées to try. Balance "For the Freezer" and "From the Freezer" dishes to make use of what you have in the freezer, what's ripe in the garden, or what needs to be used up from the refrigerator. A bargain buy on tomatoes might yield spaghetti sauce, a beef-vegetable soup, and a pan of lasagna for the freezer—all for pennies a serving.

Note when you'll need to thaw dishes for the next meal. Dacyczyn recommends getting into the habit of removing what you need from the freezer and setting it in the refrigerator to thaw the night before you need it.

Large bulk buys, such as produce, need some forethought to freeze economically. Most fruits and vegetables are easy to freeze, but they freeze best partially cooked or at least pureed, so the thawing process doesn't turn them watery and flavorless. If you have an end-of-summer trip to the farmers' market planned, also schedule a couple of weekend hours at home to prepare your bounty for freezing. Your time investment will pay off deliciously in recipe-ready ingredients throughout the winter.

Your freezer puts perishables in a holding pattern so that they're ready to use when you're ready to cook. "If you prepare the building blocks for recipes and freeze them," Roberts says, "you can spend time on the details of a meal—the extras that give dishes the best spark and flavor." Your freezer makes this possible; it'll be as easy as making ice cubes.

The Big Freeze

Most restaurant chefs rely on cooking in large batches to make their time as profitable as possible. Cooking in quantity is a basic thrift. It builds a repertoire of ready ingredients, and it saves huge amounts of hands-on time. Why cook small pots of spaghetti sauce or split-pea soup for dinner when it takes no more time to cook a double batch and freeze half? Quantity cooking allows you to stock your freezer with cooked beans, grains, pasta, soups, stews, casseroles, and sauces—the basis for many effortless meals.

Thawed and reheated frozen dishes are like brand-new meals, with none of the work. Every time you prepare a favorite recipe, ask yourself if you can make extra for next week or next month.

Or when you use only half a container of food, think of the freezer, Dacyczyn advises. For example, ½ cup of leftover fruit juice or homemade yogurt can be saved for freezer pops. Save the water you use when steaming vegetables and freeze it in a container as a nutritious addition to stews and soups. Dacyczyn keeps leftover bits of meat, cheese, and vegetables in a soup-collecting freezer bag.

McCoy calms the breakfast rush by filling her freezer with nutritious fast foods: muffins, quick breads, and other low-fat, low-cost breakfast foods baked when she has extra time.

Roberts stockpiles a selection of staples he calls freezer helpers: minced garlic, cooked chopped onions, concentrated (reduced) stocks, tomato sauces, pestos, salsas, cookie dough, and pie crust. Instead of 20 minutes of prep time, it might take Roberts only 5 minutes to have a lentil soup on the stove because he has frozen cooked onions and concentrated stock on hand.

Roberts also keeps on hand the main ingredients for quick

MAKE YOUR OWN YOGURT

Making your own yogurt takes about 10 minutes of hands-on time, and each quart saves you almost $2.50 over store-bought. For the starter, be sure to buy yogurt that has no gelatin, cornstarch, or other stabilizers. Check by scooping out a spoonful of yogurt. If the hole left by the spoon fills up with liquid within a few minutes, it is the right kind of yogurt to use. Use clean quart jars to make and store the yogurt. Provide a constant temperature by wrapping the jars in an old down jacket or quilt or by setting them on a heating pad (the kind used for backaches) and wrapping them in a blanket. We use noninstant powdered milk because it is much less expensive than instant.

3¾ quarts warm water (about 100°F)
1 cup low-fat plain yogurt made without stabilizers
4 cups noninstant powdered skim milk

If using a heating pad, set the temperature to medium and place it in a draft-free corner of the kitchen counter. Cover the pad with a towel.

Pour the water into into a large mixing bowl. Add the yogurt and the powdered milk. Whisk until the milk powder dissolves. Pour into 4 warmed 1-quart glass jars. Cap the jars.

Place the jars on the heating pad. Cover with additional towels. Let the yogurt stand for 3 to 4 hours, or until completely set. Refrigerate.

Makes 4 quarts

Per 1 cup
Calories 70
Total fat 0.3 g.
Saturated fat 0.2 g.
Cholesterol 4 mg.
Sodium 103 mg.
Fiber 0 g.

Cost per serving
17¢

freezer meals—staples like fish fillets, boneless chicken breasts, and turkey tenderloins. These freeze quickly in a single layer on a baking sheet; then they can be sandwiched between layers of freezer-quality plastic wrap, which makes it easy to pull out one or two to thaw for a superfast main dish.

Follow the advice of our experts and you'll feel healthy, wealthy, and wise knowing that a bounty of good food is always on hand. The chapters that follow give you scores of delectable recipes and tips to tap into the power of your freezer.

CHAPTER 2

APPETIZERS AND OTHER LITTLE DISHES

S tock your freezer with fast food that's also good food. Make healthy appetizers when you have extra time, to enjoy when you're most pressed for time.

Homemade freezer snacks will help you short-circuit the urge to dash out for pizza, doughnuts, or burgers. Each tasty dish from your freezer saves money and adds nutrition to your overall eating plan.

For entertaining, cooking appetizers ahead and freezing them allows you to be a guest at your own party. Wonderful morsels like Armenian Meatballs and Stuffed Mushrooms are simple to cook in quantity.

For spur-of-the-moment get-togethers, rely on frozen ingredients—like the ones in Simple Antipasto and Japanese Shrimp Salad—to whip up party snacks fast.

Team up with your freezer to create low-cost, healthy appetizers and other little dishes that make effortless party fare, snacks, and light meals.

19

Per ¼ cup
Calories 116
Total fat 0.8 g.
Saturated fat 0.1 g.
Cholesterol 2 mg.
Sodium 272 mg.
Fiber 1.2 g.

Cost per serving

36¢

KITCHEN TIP

Basil is great to have on hand in winter and reduces the cost of this dip to 9¢ a serving. When basil is abundant and cheap, chop the leaves and pack in freezer-quality plastic bags in the freezer for up to 6 months. It's easy to remove just the amount you need from the bag.

PESTO DIP

This dip freezes well. Just pack it in a freezer-quality plastic container. Thaw overnight in the refrigerator, then process briefly in a blender or food processor to restore its creamy texture.

2   cups chopped fresh basil
1   cup dry bread crumbs
3   tablespoons grated Parmesan cheese
2   tablespoons chopped almonds
2   cloves garlic
¼   cup nonfat mayonnaise
¼   cup nonfat sour cream

❋ In a blender or food processor, combine the basil, bread crumbs, Parmesan, almonds, and garlic. Process until smooth, scraping down the sides of the container as necessary. Add the mayonnaise and sour cream. Process briefly to combine.

BLUE CHEESE BONUS

Blue cheese keeps particularly well in the freezer. When you find it on sale, buy extra and crumble it onto a baking sheet lined with wax paper. Freeze solid, then transfer the cheese to a freezer-quality container or freezer-quality plastic bag. Because it freezes in separate pieces, the frozen blue cheese is easily measured out for individual recipes.

VEGETABLE MEDLEY WITH CREAMY CURRY DIP

This colorful appetizer platter is a great example of how your freezer becomes your best friend when you're planning a party. Frozen broccoli or carrot slices are often cheaper than fresh. The curry dip can be made ahead and frozen for up to 2 months. Thaw it overnight in the refrigerator and process briefly in a blender before serving.

Vegetables

- 2 cups frozen broccoli florets, thawed
- 2 cups frozen sliced carrots, thawed
- 8 radishes, thinly sliced
- 1 celery root, peeled and thinly sliced

Dip

- 1 cup low-fat ricotta cheese
- ¼ cup sliced scallions
- 2 teaspoons mango chutney
- ½ teaspoon curry powder

✳ **To make the vegetables:** Bring a large pot of water to a boil over medium-high heat. Add the broccoli and carrots. Cook for 30 seconds, or until the broccoli is bright green. Drain in a colander. Rinse under cold water. Pat dry with paper towels. Arrange on a serving platter, leaving space in the center for the dip bowl. Surround with the radishes and celery root.

✳ **To make the dip:** In a blender or food processor, process the ricotta until smooth. Scrape down the container as needed. Transfer to a serving bowl. Add the scallions, chutney, and curry powder. Stir well to combine.

Makes 4 servings

Per serving
Calories 207
Total fat 5.8 g.
Saturated fat 3.2 g.
Cholesterol 19 mg.
Sodium 316 mg.
Fiber 6.6 g.

Cost per serving

67¢

KITCHEN TIP

Celery root, also known as celeriac, is a knobby, brown root vegetable with a pleasant crunch and a flavor that tastes like a cross between celery and parsley. Look for celery root in your supermarket, and choose firm, unblemished specimens. To prevent sliced celery root from browning, soak the slices briefly in a bowl of cold water containing 2 tablespoons vinegar.

Per tomato
Calories 16
Total fat 0.5 g.
Saturated fat 0 g.
Cholesterol 3 mg.
Sodium 30 mg.
Fiber 0.2 g.

Cost per serving

9¢

KITCHEN TIP

To freeze, pack the basil stuffing in a freezer-quality plastic container. To use, thaw overnight in the refrigerator. If it separates, process it briefly in a blender or food processor to restore its creamy texture.

CHERRY TOMATOES WITH CREAMY BASIL STUFFING

When your summer garden is bursting with fresh basil, you'll want to make extra batches of this aromatic ricotta stuffing. It will keep well in the freezer for up to 3 months. Spoon into snow peas, spread it on slices of crisp zucchini or yellow summer squash, or use it to fill cherry tomatoes as we do here.

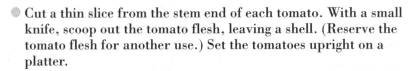

30 cherry tomatoes
1 cup low-fat ricotta cheese
¼ cup chopped fresh basil
2 cloves garlic, minced
½ teaspoon lemon juice
¼ teaspoon salt

❋ Cut a thin slice from the stem end of each tomato. With a small knife, scoop out the tomato flesh, leaving a shell. (Reserve the tomato flesh for another use.) Set the tomatoes upright on a platter.

❋ In a medium bowl, combine the ricotta, basil, garlic, lemon juice, and salt. Mix well. Spoon into the tomato shells.

SIMPLE ANTIPASTO

Restaurant antipastos can contain up to 15 grams of fat and cost $6 or more a serving. This easy version is based on frozen vegetables that are quick-cooked, then marinated in a sprightly orange dressing.

Makes 4 servings

Per serving
Calories 140
Total fat 3.8 g.
Saturated fat 1.8 g.
Cholesterol 8 mg.
Sodium 358 mg.
Fiber 6.5 g.

Cost per serving

60¢

6   leaves lettuce
2   cups frozen sliced green beans, thawed
1   cup frozen sliced carrots, thawed
1   cup frozen broccoli florets, thawed
1   cup frozen artichoke hearts, thawed
4   stalks celery, cut into 4" strips
½   cup cubed low-fat mozzarella cheese
6   pitted black olives, halved lengthwise
¼   cup orange juice
¼   cup balsamic vinegar
1   teaspoon packed brown sugar
½   teaspoon ground black pepper
¼   teaspoon salt

❋ Arrange the lettuce on a platter.

❋ Bring a large pot of water to a boil over medium-high heat. Add the beans, carrots, broccoli, and artichoke hearts. Cook for 30 seconds, or until the beans are bright green. Drain in a colander. Rinse under cold water. Pat dry with paper towels. Arrange on the lettuce. Surround with the celery, mozzarella, and olives.

❋ In a small bowl, combine the orange juice, vinegar, brown sugar, pepper, and salt. Whisk well. Drizzle over the vegetables. Let stand at room temperature for 30 minutes for the flavors to blend.

DRESS UP FROZEN VEGETABLES

For a really easy appetizer salad, marinate a selection of frozen vegetables in Parmesan-Pepper Dressing (page 264) for 30 to 40 minutes, then toss with crisp greens, chopped apples, or shredded carrots. The vegetables will marinate as they thaw. Best choices are frozen green beans, peas, carrots, and cauliflower and broccoli florets.

Per wedge
Calories 137
Total fat 4.8 g.
Saturated fat 0.9 g.
Cholesterol 0 mg.
Sodium 193 mg.
Fiber 2.5 g.

Cost per serving

33¢

KITCHEN TIP

To freeze, pack the cooled roasted vegetables tightly into freezer-quality plastic containers. To use, thaw overnight in the refrigerator.

FRENCH ROASTED VEGETABLE SANDWICHES

This easy sandwich, inspired by those served in the south of France, makes a great brown-bag lunch because it keeps for several hours without getting soggy. Cut into smaller portions, it's an out-of-the-ordinary party appetizer. Stockpile the roasted vegetables and the loaf of Italian bread separately in your freezer to thaw for quick sandwich assembly.

1   small eggplant, peeled and cut into thick slices
1   sweet red pepper, quartered
1   medium tomato, halved
1   small onion, cut into thick slices
2   tablespoons olive oil
2   teaspoons minced garlic
½   teaspoon crushed dried rosemary
1   round loaf Italian bread (8" diameter)
2   tablespoons nonfat plain yogurt
3   tablespoons balsamic vinegar
2   teaspoons grated Parmesan cheese
½   cup tightly packed spinach leaves

FOR the Freezer

✷ Preheat the oven to 400°F. Coat a large baking sheet with no-stick spray. Arrange the eggplant, peppers, tomatoes, and onions on the sheet. Brush with the oil. Sprinkle with the garlic and rosemary. Bake for 45 minutes, or until golden brown and tender.

✷ Split the bread horizontally and scoop out the interior, leaving a 1″ shell. (Reserve the bread for another use.) Spread the yogurt over the bottom of the shell, then sprinkle with the vinegar. Arrange the vegetables in the bottom of the shell. Sprinkle with the Parmesan. Top with the spinach. Place the top of the bread over the filling. Wrap tightly in plastic wrap and refrigerate for 30 minutes, or until chilled. Cut into 8 wedges.

STUFFED MUSHROOMS

Savory stuffed mushrooms are popular low-cost appetizers that can be made up to 3 months ahead and frozen. Winter sales on mushrooms lower the price to $1.20 or less a pound.

20 medium mushrooms
1 small onion, finely chopped
¼ cup frozen defatted Chicken Stock
      (page 61), thawed
2 cloves garlic, minced
1 teaspoon olive oil
½ cup soft bread crumbs
2 tablespoons chopped fresh parsley
2 teaspoons grated Parmesan cheese

❋ Preheat the broiler.

❋ Separate the stem from each mushroom cap. Coarsely chop the stems and reserve. Place the mushroom caps, round side up, on a large baking sheet. Broil 4″ from the heat for 4 minutes, or until the mushrooms are wrinkled and exude moisture. Remove from the oven and set aside to cool.

❋ In a 10″ no-stick skillet, combine the onions, stock, garlic, oil, and the reserved mushroom stems. Cook, stirring frequently, over medium-high heat for 10 minutes, or until the onions are soft but not browned. Remove from the heat. Add the bread crumbs and parsley. Stir well to combine.

❋ Pack the onion mixture into the mushroom caps. Set the mushrooms, stuffed side up, on the baking sheet. Sprinkle with the Parmesan. Broil 4″ from the heat for 5 minutes, or until golden brown.

Makes 20

Per mushroom
Calories 13
Total fat 0.4 g.
Saturated fat 0 g.
Cholesterol 0 g.
Sodium 11 mg.
Fiber 0.5 g.

Cost per serving

6¢

KITCHEN TIP

To freeze, stuff the mushrooms and place them on a tray. Put in the freezer for several hours, or until solid. Transfer to a freezer-quality plastic bag. Don't thaw the mushrooms before broiling. Just sprinkle with the Parmesan and add an extra 3 minutes to the broiling time to heat them through.

Appetizers and Other Little Dishes

Makes 12

Per cake
Calories 82
Total fat 2.3 g.
Saturated fat 1.3 g.
Cholesterol 6 mg.
Sodium 79 mg.
Fiber 1.2 g.

Cost per serving

9¢

KITCHEN TIP

To freeze, place the cornmeal cakes on a tray and put in the freezer for several hours, or until solid. Transfer to a freezer-quality plastic bag. Pack the salsa in a separate freezer-quality plastic container. Both can be frozen for up to 2 months. To use, thaw both in the refrigerator overnight. Bring the salsa to room temperature. Broil the cornmeal cakes 4" from the heat until hot.

CHEESY CORNMEAL CAKES WITH SALSA

Inspired by Italian polenta, or cooked cornmeal, these economical corn-meal cakes are as easy to make as they are to enjoy.

Salsa

- 1 small onion, minced
- 1 cup chopped tomatoes
- 2 cloves garlic, minced
- 1 tablespoon minced fresh basil
- 1 teaspoon crushed red-pepper flakes

Cornmeal Cakes

- 1¼ cups yellow cornmeal
- 1 cup cold water
- 2 tablespoons grated Parmesan cheese
- ½ cup shredded low-fat mozzarella cheese
- ½ cup shredded low-fat extra-sharp Cheddar cheese

❀ *To make the salsa:* In a small bowl, combine the onions, tomatoes, garlic, basil, and red-pepper flakes. Mix well. Let stand for 30 minutes, stirring occasionally.

❀ *To make the cornmeal cakes:* In a medium saucepan, combine the cornmeal and water; bring to a boil over medium-high heat. Cook, stirring frequently, for 12 to 15 minutes, or until the mixture is thick. Remove from the heat; stir in the Parmesan.

❀ Line a baking pan with wax paper; spoon the cornmeal mixture onto the tray and spread to ½" thickness. Cover with plastic wrap and refrigerate for 20 minutes, or until firm.

❀ Preheat the broiler. Cut the cornmeal mixture into 12 pieces; place on a large baking sheet. Coat with no-stick spray. Broil 4" from the heat for 2 minutes, or until golden brown. Turn and coat the other side with the spray. Top with the mozzarella and Cheddar. Broil for 2 minutes, or until the cheese melts. Top with the salsa.

CHILI TORTILLA STRIPS

This snack couldn't be simpler. Strips of flour tortillas are dusted with a spicy mix, then baked until golden and crisp. You can find flour tortillas on sale for as little as 99¢ per package of 6. They freeze well for up to a year.

1   teaspoon chili powder
1   teaspoon garlic powder
½   teaspoon salt
¼   teaspoon ground cumin
¼   teaspoon ground black pepper
8   frozen flour tortillas (6" diameter), thawed
2   teaspoons olive oil

❋ Preheat the oven to 400°F. Coat a large baking sheet with no-stick spray.

❋ In a small bowl, combine the chili powder, garlic powder, salt, cumin, and pepper. Mix well.

❋ Brush both sides of the tortillas with the oil. Sprinkle with the spice mixture. With kitchen scissors, cut each tortilla in half, then cut each half into 6 strips. Arrange the strips in a single layer on the prepared baking sheet. Bake in 2 batches, if necessary, to avoid crowding. Bake for 10 minutes, or until golden brown and crisp.

Makes 8 servings

Per serving
Calories 127
Total fat 3.6 g.
Saturated fat 0.8 g.
Cholesterol 0 mg.
Sodium 306 mg.
Fiber 1 g.

Cost per serving

21¢

KITCHEN TIP

The thinner the tortilla, the crisper the tortilla strip. You can substitute corn tortillas for the flour ones if you like.

SALSA SECRETS

Fresh summer salsas make great low-fat side dishes and toppings for everything from soup to fish. When sweet peppers, hot chili peppers, cilantro, and tomatoes are on sale in the summer, make extra batches of salsa. Pack it into 1-cup containers and freeze for up to 6 months. You save about 40¢ a cup over store-bought salsa.

Tex-Mex Cheese Tortilla Wedges

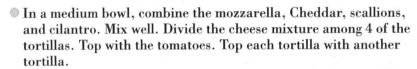

Per wedge
Calories 64
Total fat 1.4 g.
Saturated fat 0.7 g.
Cholesterol 3 mg.
Sodium 118 mg.
Fiber 0.4 g.

Cost per serving

10¢

KITCHEN TIP

Freeze the cooled cooked tortilla sandwiches whole. Layer them between sheets of freezer-quality foil, then wrap well in foil. To use, thaw overnight in the refrigerator, then reheat in the oven at 350°F for 15 minutes, or until hot. Cut into wedges; top with the yogurt or sour cream.

These are convenient to have tucked in the freezer for parties as well as for lunch or supper main dishes. This recipe makes enough to serve 4 as an entrée, accompanied by soup and a green salad.

1 cup shredded nonfat mozzarella cheese
1 cup shredded low-fat extra-sharp
     Cheddar cheese
½ cup chopped scallions
½ cup chopped fresh cilantro
8 flour tortillas (10" diameter)
1 medium tomato, thinly sliced
½ cup nonfat plain yogurt or nonfat sour cream

In a medium bowl, combine the mozzarella, Cheddar, scallions, and cilantro. Mix well. Divide the cheese mixture among 4 of the tortillas. Top with the tomatoes. Top each tortilla with another tortilla.

In a 10" no-stick skillet over medium-high heat, cook each tortilla sandwich for 3 minutes, or until golden brown. Turn and cook for 1 minute, or until the cheese melts. Transfer each to a plate. With kitchen scissors or a serrated knife, cut each tortilla sandwich into 8 wedges. Top with the yogurt or sour cream.

BLACK-BEAN CAKES
WITH SOUTHWESTERN SAUCE

Dressed with a colorful tomato sauce, these easy panfried cakes make a very pretty party dish. The black-bean cakes and the sauce can be frozen separately for up to 3 months.

Sauce

2 cups chopped tomatoes
⅓ cup tomato paste
¼ cup chopped fresh cilantro
2 cloves garlic, minced
½ teaspoon hot-pepper sauce

Bean Cakes

1 cup chopped onions
2 cloves garlic, minced
2 cups cooked black beans (page 177)
2 tablespoons grated Parmesan cheese
1 teaspoon chili powder
¼ teaspoon hot-pepper sauce
¼ cup yellow cornmeal

✸ *To make the sauce:* In a blender or food processor, combine the tomatoes, tomato paste, cilantro, garlic, and hot-pepper sauce. Process until smooth. Refrigerate.

✸ *To make the bean cakes:* Coat a 10" no-stick skillet with no-stick spray and set over medium-high heat. When the skillet is hot, add the onions and garlic. Cook and stir for 3 minutes, or until the onions are soft but not browned. Spoon into a medium bowl.

✸ Add the beans, Parmesan, chili powder, and hot-pepper sauce. Stir well to combine, then mash with a fork. With your hands, press the bean mixture into 12 cakes about ½" thick. Coat the top and bottom of each cake with the cornmeal. Place the cakes on a plate and cover. Refrigerate for 1 hour.

✸ Coat a 10" no-stick skillet with no-stick spray and set over medium-high heat. When the skillet is hot, add the cakes. Cook in batches, if necessary, to avoid crowding. Cook for 3 minutes. Turn and cook for 2 minutes, or until golden brown. Top with the sauce.

Makes 6 servings

Per serving
Calories 137
Total fat 1.6 g.
Saturated fat 0.6 g.
Cholesterol 2 mg.
Sodium 166 mg.
Fiber 6.9 g.

Cost per serving

26¢

KITCHEN TIP

To freeze, place the bean cakes on a tray and put in the freezer for several hours, or until solid. Transfer to a freezer-quality plastic bag. Pack the sauce in a freezer-quality plastic container. To use, thaw both overnight in the refrigerator. Bring the sauce to room temperature. Reheat the cakes in a covered 10" no-stick skillet over medium heat for 15 minutes, or until hot.

29

Appetizers and Other Little Dishes

Japanese Shrimp Salad

Despite its pricey reputation, shrimp can be a bargain. Peeled frozen shrimp is a frequent supermarket and warehouse store special.

Makes 4 servings

Per serving
Calories 173
Total fat 0.6 g.
Saturated fat 0.1 g.
Cholesterol 20 mg.
Sodium 196 mg.
Fiber 4.1 g.

Cost per serving

49¢

¼ cup white wine or water
2 ounces frozen uncooked peeled medium shrimp
½ cup sugar
½ cup cider vinegar
¼ teaspoon salt
2 large cucumbers, peeled, seeded, and thinly sliced
2 cups frozen cauliflower florets
2 cups frozen sliced carrots
4 leaves lettuce

* In a small saucepan, combine the wine or water and shrimp. Bring to a boil over medium-high heat. Cook, stirring, for 2 minutes, or until the shrimp turn pink and are cooked through. Check by cutting into 1 shrimp; it should be opaque in the center. Drain well and chop coarsely.

* In a large bowl, combine the sugar, vinegar, and salt. Mix well. Add the cucumbers, cauliflower, carrots, and shrimp. Toss well. Cover and refrigerate for 1 hour, stirring occasionally.

* Line 4 small plates with the lettuce and arrange the salad on top.

Shrimp Toasts

A favorite appetizer at Chinese-American restaurants, shrimp toasts are inexpensive and simple to make with frozen shrimp bought on sale. Look for shrimp in bulk bags in warehouse stores.

Makes 16

Per toast
Calories 59
Total fat 0.9 g.
Saturated fat 0 g.
Cholesterol 20 mg.
Sodium 153 mg.
Fiber 1.5 g.

Cost per serving

21¢

8 slices whole-wheat bread
8 ounces frozen peeled and cooked medium shrimp, thawed
1 egg white
1 teaspoon minced fresh ginger
½ teaspoon minced garlic
¼ teaspoon salt
¼ teaspoon ground black pepper
2 tablespoons minced scallions

- Preheat the broiler. Cut each slice of bread in half diagonally and arrange on a large baking sheet.

- In a blender or food processor, combine the shrimp, egg white, ginger, garlic, salt, and pepper. Process until smooth, scraping down the sides of the container as necessary. Transfer to a medium bowl. Stir in the scallions. Spread on the bread.

- Broil 4″ from the heat for 3 minutes, or until golden brown.

TERIYAKI CHICKEN KABOBS

Save money and keep sodium low by making your own teriyaki sauce for these easy appetizers. Big batches of the cooked kabobs can be frozen.

8   ounces chicken breasts, skinned, boned, and
        cut into 1″ cubes
⅓   cup lemon juice
2   tablespoons reduced-sodium soy sauce
2   tablespoons grated fresh ginger
1   teaspoon olive oil
1   teaspoon dried thyme
1   sweet red pepper, cut into 1″ cubes
1   small onion, cut into 1″ pieces
1   can (8 ounces) unsweetened pineapple chunks, drained

- In a large shallow nonmetal dish, combine the chicken, lemon juice, soy sauce, ginger, oil, and thyme. Stir well to combine. Cover and refrigerate for 8 hours, stirring occasionally.

- Preheat the grill or broiler. Coat the grill rack or broiler pan with no-stick spray. Thread the chicken on eight 6″ metal skewers, alternating with the peppers, onions, and pineapple. Grill or broil 4″ from the heat, basting frequently with the marinade, for 5 minutes. Turn and cook for 5 minutes, or until the vegetables are golden brown and the chicken is no longer pink in the center. Check by inserting the tip of a sharp knife into 1 cube.

KITCHEN TIP

To freeze the cooked shrimp toasts, place them on a tray and put in the freezer for a few hours. Transfer to a freezer-quality plastic bag. To use, thaw overnight in the refrigerator. Place on a large baking sheet. Reheat in a 400°F oven for 10 minutes.

Makes 8

Per kabob
Calories 59
Total fat 1.3 g.
Saturated fat 0.3 g.
Cholesterol 16 mg.
Sodium 165 mg.
Fiber 0.4 g.

Cost per serving

34¢

KITCHEN TIP

To freeze the kabobs, cool the chicken and remove it from the skewers. Place on a tray and freeze for a few hours. Transfer to freezer-quality plastic bags. To use, thaw overnight in the refrigerator. Broil 4″ from the heat for 2 to 3 minutes.

Appetizers
and Other Little Dishes

Per serving
Calories 149
Total fat 3.8 g.
Saturated fat 0.6 g.
Cholesterol 32 mg.
Sodium 220 mg.
Fiber 0.7 g.

Cost per serving

59¢

KITCHEN TIP

To freeze the baked nuggets, place them on a tray and put in the freezer for several hours, or until solid. Transfer to a freezer-quality plastic bag. Pack the salsa in a freezer-quality plastic container. To use, thaw both overnight in the refrigerator. Place the nuggets on a large baking sheet, cover wth foil to keep them from drying out, and reheat at 350°F for 10 minutes.

CHICKEN NUGGETS WITH ORANGE SALSA

Having these cooked chicken nuggets ready in the freezer gives you easy party fare or a popular main course for children's birthday parties. Each time you make the recipe, you'll save $1 a serving over packaged frozen chicken nuggets.

Nuggets

¼ cup honey
¼ cup Dijon mustard
1 tablespoon lemon juice
1 tablespoon curry powder
1 pound chicken breasts, skinned, boned, and cut into 1" cubes

Salsa

2 navel oranges
1 small jalapeño pepper, seeded and minced (wear plastic gloves when handling)
2 tablespoons packed brown sugar
2 tablespoons chopped fresh cilantro
1 tablespoon oil

※ *To make the nuggets:* Preheat the oven to 350°F.

※ In a small saucepan, combine the honey, mustard, lemon juice, and curry powder. Mix well. Bring to a boil over medium-high heat. Pour into a medium bowl. Add the chicken and toss well to coat.

※ Coat a large baking sheet with no-stick spray. Arrange the chicken on the baking sheet in a single layer. Bake for 25 minutes, or until the chicken is no longer pink in the center. Check by inserting the tip of a sharp knife into 1 cube.

※ *To make the salsa:* Peel 1 orange and chop it; place in a medium bowl. Juice the other orange and add to the bowl. Add the peppers, brown sugar, cilantro, and oil. Stir well to combine. Let stand at room temperature for 20 minutes, stirring occasionally. Serve with the nuggets.

CHICKEN DRUMMETTES WITH BLUE CHEESE DIP

Buy frozen chicken drummettes (chicken wings with the tips removed) in bulk at your local warehouse store and save at least 50¢ a pound over supermarket prices. You can marinate, bake, and freeze the chicken for up to 3 months.

Chicken Drumettes

1 pound frozen chicken drummettes (24), thawed
¼ cup hot-pepper sauce
3 tablespoons cider vinegar
1 teaspoon oil

Blue Cheese Dip

1 cup low-fat ricotta cheese
½ cup nonfat plain yogurt
1 clove garlic, minced
2 tablespoons crumbled blue cheese
¼ teaspoon paprika

❋ *To make the chicken drummettes:* Remove as much skin as possible from the chicken. In a large shallow nonmetal baking dish, combine the chicken, hot-pepper sauce, vinegar, and oil. Toss to combine. Cover and refrigerate for 8 hours, turning occasionally.

❋ Preheat the oven to 350°F. Line a large baking pan with foil. Transfer the chicken to the pan. Discard any marinade that doesn't cling to the chicken. Bake for 30 minutes, or until the chicken is no longer pink in the center. Check by inserting the tip of a sharp knife into the thickest part of 1 drummette.

❋ *To make the blue cheese dip:* In a blender or food processor, combine the ricotta, yogurt, garlic, and 1 tablespoon of the blue cheese. Process until smooth. Spoon into a serving bowl. Stir in the remaining 1 tablespoon blue cheese. Sprinkle with the paprika. Cover and refrigerate for 20 minutes, or until the flavors blend.

Makes 24

Per drumette
Calories 65
Total fat 2.9 g.
Saturated fat 1.2 g.
Cholesterol 22 mg.
Sodium 47 mg.
Fiber 0 g.

Cost per serving

16¢

KITCHEN TIP

To freeze, place the cooled cooked drummettes on a tray. Put in the freezer for several hours, or until solid. Transfer to a freezer-quality plastic bag. Pack the dip in a freezer-quality plastic container. To use, thaw both overnight in the refrigerator. Place the drummettes on a large baking sheet and reheat at 350°F for 15 minutes, or until hot. Process the dip briefly in a blender or food processor until creamy.

Appetizers and Other Little Dishes

Per serving
Calories 47
Total fat 0.6 g.
Saturated fat 0.2 g.
Cholesterol 8 mg.
Sodium 43 mg.
Fiber 0.6 g.

Cost per serving

14¢

KITCHEN TIP

To freeze, pack
the cooled
cooked beef in
a freezer-quality
plastic container.
Freeze for up to
3 months. To use,
thaw overnight
in the refrigerator.
Microwave on
high power for
5 minutes, or
until hot.

THAI BEEF ON ORANGE SLICES

This traditional Thai snack—spicy ground beef served on orange slices—makes an unusual and refreshing appetizer. For authentic flavor, substitute Thai fish sauce for the soy sauce. In Asian stores or most supermarkets, fish sauce costs less than $1 per 16-ounce bottle.

4   large navel oranges
½   small onion, minced
3   cloves garlic, minced
1   tablespoon water
½   teaspoon oil
8   ounces extra-lean ground round beef
1   small jalapeño pepper, seeded and minced
     (wear plastic gloves when handling)
2   tablespoons packed brown sugar
1   tablespoon reduced-sodium soy sauce
½   teaspoon cider vinegar
2   tablespoons minced fresh cilantro

✸ Peel the oranges. Cut each crosswise into 4 thick slices. Arrange on a platter.

✸ In a 10″ no-stick skillet, combine the onions, garlic, water, and oil. Cook, stirring frequently, over medium-high heat for 3 minutes, or until the onions are soft but not browned. Add the beef. Cook and stir for 5 minutes, or until the beef is no longer pink. Add the peppers, brown sugar, soy sauce, and vinegar. Cook, stirring frequently, for 3 minutes, or until the liquid evaporates. Add the cilantro. Stir well to combine.

✸ Top each orange slice with the beef mixture.

Armenian Meatballs

Bursting with flavor, these bite-size appetizers are simple to make and surprisingly low in cost and fat. The cooked meatballs and sauce can be frozen for up to 4 months and make excellent holiday party fare. For parties, serve the meatballs and sauce in a fondue pot or chafing dish with toothpicks alongside.

8 ounces extra-lean ground round beef
¼ cup soft bread crumbs
2 tablespoons water
1 clove garlic, minced
½ teaspoon ground cumin
¼ teaspoon salt
¼ teaspoon ground black pepper
1 can (14 ounces) reduced-sodium whole tomatoes, chopped (with juice)
1 can (8 ounces) reduced-sodium tomato sauce

✹ In a medium bowl, combine the beef, bread crumbs, water, garlic, cumin, salt, and pepper. Mix well. Form into 1" balls. Place in a 10" no-stick skillet. Add the tomatoes (with juice) and tomato sauce. Bring to a boil over medium-high heat. Reduce the heat to medium. Cover and cook for 25 minutes, or until the meatballs are no longer pink in the center. Check by inserting the tip of a sharp knife into the center of 1 meatball.

Makes 16

Per meatball
Calories 29
Total fat 0.6 g.
Saturated fat 0.2 g.
Cholesterol 8 mg.
Sodium 48 mg.
Fiber 0.7 g.

Cost per serving

6¢

Kitchen Tip

To freeze, pack the cooled cooked meatballs and sauce in a freezer-quality plastic container. To use, thaw overnight in the refrigerator. Reheat in a 10" no-stick skillet, covered, over medium heat for 10 minutes, or until bubbling.

Appetizers
and Other Little Dishes

Per serving
Calories 285
Total fat 7.1 g.
Saturated fat 3 g.
Cholesterol 30 mg.
Sodium 230 mg.
Fiber 2.1 g.

Cost per serving

82¢

KITCHEN TIP

To freeze, pack the cooled cooked steak in a freezer-quality plastic bag or container. Pack the onions separately and freeze. To use, thaw both overnight in the refrigerator. Place the steak on a broiler pan lightly coated with no-stick spray. Broil 4" from the heat for 2 minutes, or until hot. Reheat the onions in a 10" no-stick skillet.

OPEN-FACED PEPPERED STEAK SANDWICHES

Lean flank steak is the most economical choice for these marinated steak sandwiches. You can buy it on sale for as little as $4.50 a pound, then freeze it for up to 4 months. This hearty appetizer also makes a great lunch for 4.

1   pound flank steak, trimmed of fat
½   cup balsamic vinegar
1   teaspoon ground black pepper
¼   cup frozen defatted Chicken Stock
      (page 61), thawed
2   cups chopped onions
1   loaf French bread (about 8 ounces)
1   teaspoon olive oil
1   tablespoon low-fat blue cheese, crumbled
2   teaspoons minced garlic

❀ Place the steak in a large shallow nonmetal dish and drizzle with the vinegar. Cover and refrigerate overnight, turning occasionally.

❀ Remove the steak from the vinegar and pat dry. Rub with the pepper.

❀ Bring the stock to a boil in a 10" no-stick skillet over medium-high heat. Add the onions. Cook and stir for 10 minutes, or until soft but not browned. Increase the heat to high. Cook and stir for 3 minutes, or until golden brown. Transfer to a plate.

❀ Reduce the heat to medium-high. Add the steak. Cook for 7 minutes, or until brown. Turn and cook for 10 minutes, or until no longer pink in the center. Check by inserting the tip of a sharp knife into the center. Transfer to another plate. Let stand for 5 minutes. Cut into thin crosswise slices.

❀ Preheat the oven to 400°F. Split the bread in half lengthwise. Brush the cut sides with the oil. Place the bread, cut side up, on a baking sheet. Bake for 5 minutes, or until golden brown. Cut each half into 4 pieces. Top with the blue cheese, garlic, onions, and steak.

POP GOES THE FREEZER

Earl Proulx, *Yankee* magazine fix-it columnist and author of *Yankee Magazine's Make It Last!*, says, "If you want to make sure that your popcorn kernels pop, store them in the freezer." It protects them from moisture loss. This tasty TV snack saves you 25¢ a cup over store-bought flavored low-fat popcorn. Just pop the kernels straight from the freezer; there's no need to thaw.

BARBECUED POPCORN

- 2 tablespoons molasses
- 2 tablespoons reduced-sodium barbecue sauce
- 1 tablespoon reduced-sodium ketchup
- ½ teaspoon paprika
- ½ teaspoon garlic powder
- 8 cups air-popped popcorn

In a small bowl, combine the molasses, barbecue sauce, ketchup, paprika, and garlic powder. Mix well.

Place the popcorn in a large bowl. Drizzle with the molasses mixture. Toss well to coat.

Preheat the oven to 200°F. Line a large baking sheet with foil and coat with no-stick spray. Spread the popcorn mixture on the sheet. Bake for 20 minutes. Turn off the oven and let the popcorn cool in the oven for 30 minutes, or until crisp.

Makes 8 cups

Per 1 cup
Calories 50
Total fat 0 g.
Saturated fat 0 g.
Cholesterol 0 mg.
Sodium 34 mg.
Fiber 1.3 g.

Cost per serving

5¢

Appetizers and Other Little Dishes

CHAPTER 3

SOUPS, STEWS, AND CHOWDERS

S oup is a freezer's dream date. Most soups hibernate happily in the freezer for 4 to 6 months with no loss of flavor. In fact, some seasonings mellow in the freezer, making soups taste even better when reheated. If freezing softens the stronger spices in chili, chowder, or stew, simply add a pinch more before serving.

For "free" soup anytime, freeze small amounts of leftover cooked meat, vegetables, grains, and beans or even grated cheese, in a large plastic container. You don't even have to thaw the ingredients. When they add up to a full container, throw them into a pot, cover with stock, and simmer until the flavors blend. Serve as a chunky soup or puree in a blender or food processor with a little low-fat milk, low-fat plain yogurt, or nonfat sour cream.

You'll never be hungry again with soup in the freezer. It adds nutrition and variety to your meals for just pennies.

Per 1 cup
Calories 84
Total fat 2.5 g.
Saturated fat 0.9 g.
Cholesterol 3 mg.
Sodium 126 mg.
Fiber 1.7 g.

Cost per serving

45¢

KITCHEN TIP

To freeze, pack
the cooled soup
in a freezer-quality
plastic container.
To use, thaw
overnight in the
refrigerator.
Transfer to
a saucepan. Cover
and cook, stirring
frequently, over
low heat for
15 minutes, or
until hot.

BASIL-VEGETABLE SOUP

Summer in a bowl for less than 50¢ a serving, this hearty pesto-flavored soup takes advantage of seasonal sales on fresh basil and vegetables. You can freeze it for winter meals. Serve it with a green salad and crusty brown bread for a satisfying lunch.

1   cup chopped onions
1   teaspoon oil
1   medium tomato, chopped
1   small zucchini or yellow summer squash,
      chopped
⅓   cup diced green beans
¼   cup diced celery
4   cups frozen defatted Chicken Stock
      (page 61), thawed
¼   cup chopped fresh parsley
¼   cup broken spaghetti
1   tablespoon reduced-sodium tomato paste
½   cup chopped fresh basil
¼   cup dry bread crumbs
¼   cup grated Parmesan cheese
3   cloves garlic, minced

❋ In a Dutch oven, combine the onions and oil. Cook, stirring, over medium-high heat for 10 minutes, or until soft but not browned. Add the tomatoes, zucchini or squash, beans, celery, and 1 cup of the stock. Cook, stirring, for 10 minutes, or until soft.

❋ Add the parsley, spaghetti, tomato paste, and the remaining 3 cups stock. Bring to a boil. Cook, stirring occasionally, for 15 minutes, or until the spaghetti is tender.

❋ In a blender or food processor, combine the basil, bread crumbs, Parmesan, and garlic. Process until smooth. Stir into the soup.

CHINESE CHICKEN SOUP

Serve this clear soup over cooked rice, Chinese-style, for a warming one-dish meal. The tofu boosts the protein profile while keeping the saturated fat low, and it costs only 75¢ a pound.

- 8 ounces frozen skinned and boned chicken breasts, cut into 1" cubes
- 1 tablespoon reduced-sodium soy sauce
- ½ teaspoon sugar
- ½ teaspoon ground black pepper
- 1 cup frozen chopped onions
- 1 cup frozen sliced zucchini or yellow summer squash
- ½ cup frozen sliced carrots
- 1 teaspoon dark sesame oil
- 6 cups frozen defatted Chicken Stock (page 61), thawed
- 8 ounces reduced-fat firm tofu, cut into ½" cubes
- ½ cup minced fresh cilantro
- ⅓ cup chopped scallions
- 1 teaspoon lime juice

❁ In a shallow nonmetal dish, combine the chicken, soy sauce, sugar, and pepper. Toss to coat. Let stand for 10 minutes, stirring occasionally.

❁ In a Dutch oven, combine the onions, zucchini or squash, carrots, and oil. Cook, stirring, over medium-high heat for 5 minutes, or until the onions are soft but not browned. Add the stock; bring to a boil.

❁ Reduce the heat to medium. Cover and cook, stirring occasionally, for 20 minutes.

❁ Add the chicken and marinade. Cook for 5 to 7 minutes, or until the chicken is no longer pink in the center. Check by inserting the tip of a sharp knife into 1 cube. Add the tofu, cilantro, and scallions. Cook for 1 minute, or until the tofu is hot. Add the lime juice. Stir well to combine.

Makes 6 cups

Per 1 cup
Calories 94
Total fat 2.2 g.
Saturated fat 0.4 g.
Cholesterol 22 mg.
Sodium 163 mg.
Fiber 1.6 g.

Cost per serving

64¢

KITCHEN TIP

Get full value from any bunch of herbs you purchase. After using the amount you need for a recipe, chop and freeze the remaining leaves in a small freezer-quality plastic bag. Cilantro and other herbs will remain green and flavorful. Use the herbs while still frozen.

KITCHEN TIP

To freeze, pack the cooled soup in a freezer-quality plastic container. To use, thaw overnight in the refrigerator. Transfer to a saucepan. Cover and cook, stirring frequently, over low heat for 15 minutes, or until hot.

CITRUSY RED-PEPPER SOUP

Roasting peppers prior to freezing helps them retain their intense flavor. Make this delightful soup when peppers are abundant in your garden or are 4 for $1 at the local farmers' market.

2 large sweet red peppers, halved and seeded
2 small onions, minced
¼ cup diced carrots
2 tablespoons canned diced mild green
      chili peppers
2 cloves garlic, minced
2 teaspoons olive oil
4 cups frozen defatted Chicken Stock
      (page 61), thawed
1 large navel orange, peeled and chopped
2 tablespoons chopped fresh parsley or cilantro
½ teaspoon ground red pepper
½ teaspoon salt
¼ teaspoon ground black pepper
½ cup low-fat sour cream

❋ Preheat the broiler. Line a broiler pan with foil. Place the sweet red peppers, cut side down, on the pan and broil 4" from the heat for 15 minutes, or until blackened. Place the peppers in a brown paper bag and roll the top to seal. Let stand for 20 minutes. Remove the peppers from the bag. Peel off the blackened skin and discard. Set the peppers aside.

❋ In a Dutch oven, combine the onions, carrots, chili peppers, garlic, and oil. Cook, stirring, over medium-high heat for 10 minutes, or until the onions are soft but not browned. Add the stock, oranges, parsley or cilantro, ground red pepper, salt, and black pepper. Bring to a boil. Reduce the heat to medium. Cover and cook for 10 minutes, or until the carrots are very soft.

❋ Let the soup cool slightly. Add the red peppers and sour cream. Process in a blender or food processor until smooth. Return to the pot and heat through.

CREAMY POTATO SOUP

Potatoes can present a challenge for freezing because they absorb water and lose texture. One solution is to puree the potatoes before freezing, as in this silken soup. If freezing this soup, add the yogurt and dillweed just before serving.

3   cups diced potatoes
2   cups chopped leeks
½   cup apple juice
3   cloves garlic, minced
4   cups frozen defatted Chicken Stock
      (page 61), thawed
½   teaspoon ground nutmeg
½   teaspoon ground black pepper
¼   teaspoon salt
½   cup nonfat plain yogurt
¼   teaspoon dried dillweed

❋ In a Dutch oven, combine the potatoes, leeks, apple juice, and garlic. Cook, stirring, over medium-high heat for 10 minutes, or until the leeks are soft but not browned. Add the stock. Bring to a boil. Reduce the heat to medium. Cook, stirring occasionally, for 30 minutes, or until the potatoes are very soft when pierced with a sharp knife.

❋ Add the nutmeg, pepper, and salt. Process in a blender or food processor until smooth. Return to the pot and heat through. Top each serving with a dollop of yogurt. Sprinkle with the dillweed.

THICK AND CREAMY

Flour-thickened soups tend to separate in the freezer, so try thickening your freezer-bound soups with starchy bean or vegetable purees. One cup of puree adds body to about 3 cups of a brothlike soup. Try pureed cooked beans, potatoes, carrots, or winter squash. You'll cut calories and fat by eliminating the butter and oil that normally accompany flour thickeners.

Per 1 cup
Calories 104
Total fat 0.3 g.
Saturated fat 0.1 g.
Cholesterol 0 mg.
Sodium 114 mg.
Fiber 2.4 g.

Cost per serving

35¢

KITCHEN TIP

To freeze, pack the cooled soup in a freezer-quality plastic container. To use, thaw overnight in the refrigerator. If it separates after thawing, process in a blender or food processor until smooth. Pour into a microwaveable container. Cover and microwave on high power for 2 to 3 minutes, or until hot.

Soups, Stews, and Chowders

Italian Chickpea Soup

Robust chickpeas are great soup thickeners. At 12¢ a cup cooked, they're easy on the budget, too. You can freeze this hearty soup for up to 6 months, but add the sour cream right before serving.

2½ cups dried chickpeas
6 cups water
1 large onion, chopped
1 cup apple juice
3 cloves garlic, minced
3 stalks celery, chopped
½ cup sliced carrots
1 teaspoon dried marjoram
1 teaspoon ground black pepper
½ teaspoon salt
¼ teaspoon ground coriander
6 cups frozen defatted Chicken Stock (page 61), thawed
1 tablespoon balsamic vinegar
½ cup low-fat sour cream

* In a Dutch oven, combine the chickpeas and water. Bring to a boil over medium-high heat. Reduce the heat to medium. Cover and cook for 1½ hours, or until almost tender. Drain well and set aside.

* Add the onions, apple juice, and garlic to the pot. Bring to a boil over medium-high heat. Cook, stirring, for 5 minutes, or until the onions are soft but not browned. Add the celery, carrots, marjoram, pepper, salt, and coriander. Cook, stirring, for 5 minutes. Add the stock and reserved chickpeas. Bring to a boil.

* Reduce the heat to medium. Cook for 25 minutes, or until the chickpeas are very soft. Add the vinegar. Stir well to combine. Top each serving with a dollop of the sour cream.

Makes 8 cups

Per 1 cup
Calories 260
Total fat 4.5 g.
Saturated fat 1.1 g.
Cholesterol 5 mg.
Sodium 169 mg.
Fiber 7 g.

Cost per serving

21¢

KITCHEN TIP

To freeze, pack the cooled soup in a freezer-quality plastic container. To use, thaw overnight in the refrigerator. Transfer to a saucepan. Cover and cook, stirring frequently, over low heat for 15 minutes, or until hot.

Lentil and Escarole Soup

Dried lentils cost less than 10¢ a cup, but they pack plenty of low-fat protein into dishes like this Mediterranean soup. Traditionally made in spring, it uses escarole, a bitter lettuce. You can substitute spinach or any other leafy green that you have on hand. If freezing this soup, add the croutons and Parmesan when serving.

1   medium onion, chopped
½   cup minced carrots
¼   cup balsamic vinegar
2   cloves garlic, minced
1   teaspoon oil
5   cups frozen defatted Chicken Stock
        (page 61), thawed
1   can (8 ounces) reduced-sodium whole
        tomatoes, chopped (with juice)
1   cup dried lentils
1   bay leaf
2   cups chopped escarole
½   teaspoon ground black pepper
¼   teaspoon salt
2   slices Italian bread, cut into 1" cubes
¼   cup grated Parmesan cheese

✽ In a Dutch oven, combine the onions, carrots, vinegar, garlic, and oil. Cook, stirring, over medium-high heat for 5 minutes, or until the vegetables soften slightly. Add the stock, tomatoes, lentils, and bay leaf. Bring to a boil.

✽ Reduce the heat to medium and cook, stirring occasionally, for 40 minutes, or until the lentils are tender. Add the escarole, pepper, and salt. Stir well to combine. Cook for 5 minutes. Remove and discard the bay leaf.

✽ Preheat the oven to 425°F. Place the bread on a large baking sheet. Bake, turning occasionally, for 7 minutes, or until golden brown. Top each serving of soup with croutons. Sprinkle with Parmesan.

Makes 7 cups

Per 1 cup
Calories 159
Total fat 2.5 g.
Saturated fat 0.9 g.
Cholesterol 2 mg.
Sodium 213 mg.
Fiber 5 g.

Cost per serving

38¢

Kitchen Tip

To freeze, pack the cooled soup in a freezer-quality plastic container. To use, thaw overnight in the refrigerator. Transfer to a saucepan. Cover and cook, stirring frequently, over low heat for 15 minutes, or until hot.

Per 1 cup
Calories 143
Total fat 1 g.
Saturated fat 0.2 g.
Cholesterol 0 mg.
Sodium 167 mg.
Fiber 5.9 g.

Cost per serving

65¢

KITCHEN TIP

*Get full value from
any bunch of herbs
you purchase.
After using the
amount you need
for a recipe, chop
and freeze the
remaining leaves in
a small freezer-
quality plastic bag.
Mint and other
herbs will remain
green and flavorful.
Use the herbs
while still frozen.*

MINTED PEA SOUP

You can make this cool, pale green soup anytime with frozen peas. It's at its very best, though, prepared with locally grown fresh peas in June and July, when the price is less than 50¢ a pound.

1   cup frozen chopped onions
⅓   cup apple juice
1   teaspoon olive oil
2   cloves garlic, minced
3   cups frozen defatted Chicken Stock
      (page 61), thawed
1   cup frozen diced potatoes
2   cups frozen peas
¼   cup chopped fresh mint
½   teaspoon ground black pepper
¼   teaspoon salt
¼   cup nonfat plain yogurt

* In a Dutch oven, combine the onions, apple juice, and oil. Cook, stirring, over medium-high heat for 5 minutes, or until the onions are soft but not browned. Add the garlic. Cook, stirring, for 1 minute.

* Add the stock and potatoes. Bring to a boil. Cook for 10 minutes. Add the peas and mint. Cook for 3 minutes, or until the peas are bright green.

* Process in a blender or food processor until very smooth. Pour into a large bowl. Add the pepper and salt. Stir well. Cover and refrigerate for 2 hours, or until the soup is very cold. Top each serving with a dollop of the yogurt.

GARLIC GOODNESS

Roasted garlic puree is a timesaving
staple to store in the freezer. Drop spoonfuls
of puree onto a baking sheet lined with wax paper,
freeze solid, then transfer to a plastic bag. Add a frozen
spoonful to a soup or stew. Garlic puree can be frozen for up
to 1 year without losing its sweet and potent flavor.

ROASTED GARLIC SOUP WITH TURKEY

Roasted garlic adds velvety texture and a sweet garlic flavor to soup for pennies a serving. When garlic is in season in late spring and early summer, roast extra and freeze the puree for up to 1 year.

4 whole heads garlic
1 teaspoon olive oil
4 cups frozen defatted Chicken Stock
    (page 61), thawed
2 medium carrots, sliced
1 cup whole kernel corn
1 small onion, sliced
4 ounces turkey cutlets, chopped
½ cup minced fresh parsley
¼ teaspoon salt
¼ teaspoon ground black pepper

❋ Preheat the oven to 350°F. With a sharp knife, slice the top ¼" off each head of garlic; discard the tops. Lightly brush the heads with the oil. Place in a shallow baking dish. Cover with foil.

❋ Bake for 55 to 60 minutes. Remove the foil and bake for 10 minutes, or until the garlic skin is browned and the interior is very soft. Set aside until cool enough to handle. Squeeze the cloves to release the roasted garlic; discard the skin.

❋ In a blender or food processor, combine the garlic and 1 cup of the stock. Process until smooth.

❋ In a Dutch oven, combine the carrots, corn, onions, and 1 cup of the remaining stock. Cook, stirring, over medium-high heat for 10 minutes, or until the onions are soft but not browned. Add the turkey, garlic mixture, and the remaining 2 cups stock. Bring to a boil. Cook, stirring occasionally, for 20 minutes, or until the turkey is no longer pink in the center. Check by inserting the tip of a sharp knife into 1 piece. Stir in the parsley, salt, and pepper.

Per 1 cup
Calories 104
Total fat 1.5 g.
Saturated fat 0 g.
Cholesterol 13 mg.
Sodium 124 mg.
Fiber 2.2 g.

Cost per serving

48¢

KITCHEN TIP

To freeze, pack the cooled soup in a freezer-quality plastic container. To use, thaw overnight in the refrigerator. Transfer to a saucepan. Cover and cook, stirring frequently, over low heat for 15 minutes, or until hot.

KITCHEN TIP

Get full value from
any bunch of herbs
you purchase.
After using the
amount you need
for a recipe, chop
and freeze the
remaining leaves in
a small freezer-
quality plastic bag.
Parsley and other
herbs will remain
green and flavorful.
Use the herbs
while still frozen.

SPICED CREAM OF CARROT SOUP

To reduce the fat in this velvety soup, we replaced all the cream with low-fat buttermilk and intensified the flavor with plenty of spices. This soup freezes well; just puree it before serving if it separates.

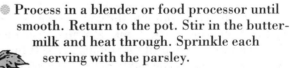

¼ cup apple juice
3 cups frozen chopped carrots
1 cup frozen chopped onions
½ cup frozen diced potatoes
2 cloves garlic, minced
1 teaspoon ground cumin
½ teaspoon ground ginger
½ teaspoon ground red pepper
½ teaspoon ground cinnamon
4 cups frozen defatted Chicken Stock
    (page 61), thawed
1½ cups low-fat buttermilk
¼ cup chopped fresh parsley

❋ Bring the apple juice to a boil in a Dutch oven over medium-high heat. Add the carrots, onions, potatoes, and garlic. Cook, stirring, for 5 minutes, or until the onions are soft but not browned. Add the cumin, ginger, pepper, and cinnamon. Cook, stirring, for 1 minute.

❋ Add the stock. Bring to a boil. Reduce the heat to medium. Cook, stirring occasionally, for 30 minutes.

❋ Process in a blender or food processor until smooth. Return to the pot. Stir in the buttermilk and heat through. Sprinkle each serving with the parsley.

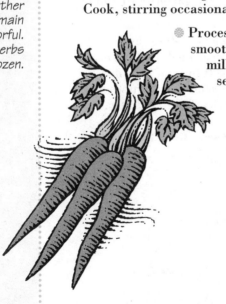

SWEET-HEAT TOMATO SOUP

This easy soup saves you 10¢ per serving over canned tomato soup, and it's much lower in salt and fat. The flavors blend even better after it's been frozen, so it makes an almost instant light supper.

1 small sweet red pepper, chopped
1 small onion, chopped
3 cloves garlic, minced
1 teaspoon olive oil
4 cups frozen defatted Chicken Stock
     (page 61), thawed
3 cups chopped tomatoes
1 small jalapeño pepper, seeded and minced
     (wear plastic gloves when handling)
1 tablespoon honey
¼ cup low-fat sour cream

✳ In a Dutch oven, combine the red peppers, onions, garlic, oil, and ½ cup of the stock. Cook, stirring, over medium-high heat for 5 minutes, or until the onions are soft but not browned. Add the tomatoes, jalapeño peppers, honey, and the remaining 3½ cups stock. Bring to a boil.

✳ Reduce the heat to medium and cook, stirring occasionally, for 25 minutes, or until thick.

✳ Process in a blender or food processor until smooth. Return to the pot. Add the sour cream and stir well. Heat through but do not boil.

HOW SWEET IT IS

Cooking soup vegetables in a small amount of oil without browning them (called sweating) brings out the sweet side of aromatics, such as onions, garlic, and celery. Because vegetables are 90 percent water, sweating evaporates some of this liquid and intensifies their flavor. Most soup recipes in this book call for cooking the vegetables without browning before adding the stock. This initial slow-cooking extracts so much flavor that you often don't need to season the soup with salt and pepper. You can sweat big batches of soup vegetables to freeze and use as you need them.

Makes 6 cups

Per 1 cup
Calories 61
Total fat 1.8 g.
Saturated fat 0.6 g.
Cholesterol 3 mg.
Sodium 21 mg.
Fiber 1.2 g.

Cost per serving

36¢

KITCHEN TIP

To freeze, pack the cooled soup in a freezer-quality plastic container. To use, thaw overnight in the refrigerator. Transfer to a saucepan. Cover and cook, stirring frequently, over low heat for 15 minutes, or until hot.

VEGETABLE AND ORZO SOUP

Makes 6 cups

Per 1 cup
Calories 258
Total fat 3.6 g.
Saturated fat 1.3 g.
Cholesterol 4 mg.
Sodium 177 mg.
Fiber 5.5 g.

Cost per serving

49¢

KITCHEN TIP

To freeze, pack the cooled soup in a freezer-quality plastic container. To use, thaw overnight in the refrigerator. Transfer to a saucepan. Cover and cook, stirring frequently, over low heat for 15 minutes, or until hot.

Orzo is a tiny rice-shaped pasta that is truly at home in hearty soups like this bowl of vegetables and garden herbs. Freezing does this soup nothing but good because it gets even more flavorful as the seasonings blend. Add the Parmesan after reheating if you freeze the soup.

1   medium onion, chopped
1   cup chopped carrots
1   cup whole kernel corn
1   small zucchini or yellow summer squash, diced
⅓   cup balsamic vinegar
1   teaspoon olive oil
4   cups frozen defatted Chicken Stock (page 61), thawed
1   cup orzo pasta
1   cup cooked navy beans (page 177)
2   cloves garlic, minced
¼   cup minced fresh parsley
1   tablespoon honey or sugar
½   teaspoon ground black pepper
⅛   teaspoon salt
⅓   cup grated Parmesan cheese

✱ In a Dutch oven, combine the onions, carrots, corn, zucchini or squash, vinegar, and oil. Cook, stirring, over medium-high heat for 5 minutes, or until the onions are soft but not browned. Add the stock, pasta, beans, and garlic. Bring to a boil.

✱ Reduce the heat to medium. Cook for 15 minutes, or until the pasta is tender.

✱ Add the parsley, honey or sugar, pepper, and salt. Stir well to combine. Sprinkle each serving with the Parmesan.

CORN AND BEEF STEW

This midwestern favorite allows you to make lean beef even healthier with plenty of potatoes, carrots, and corn. Serve it for lunch or supper with whole-grain bread and salad.

1 pound lean boned beef chuck roast,
    trimmed of fat and cut into 1" pieces
1 small jalapeño pepper, seeded and minced
    (wear plastic gloves when handling)
3 cloves garlic, minced
2 tablespoons all-purpose flour
5 cups frozen defatted Chicken Stock
    (page 61), thawed
3 russet potatoes, cubed
1 large onion, thinly sliced
1 cup whole kernel corn
1 cup diced sweet red peppers
³/₄ teaspoon ground black pepper

❉ Coat a Dutch oven with no-stick spray and place over medium-high heat until hot. Add the beef, jalapeño peppers, and garlic. Cook, stirring, for 5 minutes, or until the beef is browned. Add the flour. Cook, stirring, for 2 minutes.

❉ Add the stock, potatoes, onions, corn, and red peppers. Bring to a boil. Reduce the heat to medium. Cover and cook for 25 minutes, or until the stew is thick. Add the black pepper. Stir well to combine.

Makes 4 cups

Per 1 cup
Calories 348
Total fat 5.7 g.
Saturated fat 1.9 g.
Cholesterol 79 mg.
Sodium 71 mg.
Fiber 4.6 g.

Cost per serving

$1.09

KITCHEN TIP

To freeze, pack the cooled soup in a freezer-quality plastic container. To use, thaw overnight in the refrigerator. Transfer to a saucepan. Cover and cook, stirring frequently, over low heat for 15 minutes, or until hot.

Per 1 cup
Calories 317
Total fat 8 g.
Saturated fat 2.6 g.
Cholesterol 105 mg.
Sodium 209 mg.
Fiber 3.5 g.

Cost per serving

40¢

KITCHEN TIP

To freeze, pack
the cooled soup
in a freezer-
quality plastic
container. To use,
thaw overnight in
the refrigerator.
Transfer to a
saucepan. Cover
and cook, stirring
frequently, over
low heat for
15 minutes, or
until hot.

IRISH BEEF STEW

Potatoes and carrots are the hearty foundation of this warming stew. To cut fat but not flavor, we replaced the traditional lamb with lean beef.

3   tablespoons all-purpose flour
3   pounds lean boned beef chuck roast,
      trimmed of fat and cut into 1" pieces
½   cup apple juice
2   teaspoons olive oil
1   cup frozen defatted Chicken Stock
      (page 61), thawed
3   cups chopped onions
1   tablespoon minced garlic
1   can (28 ounces) reduced-sodium
      whole tomatoes, chopped (with juice)
3   cups cubed red potatoes
½   cup chopped carrots
½   teaspoon salt
½   teaspoon ground black pepper
¼   teaspoon ground red pepper

✺ Place the flour in a shallow dish. Dredge the beef in the flour, coating all sides.

✺ In a Dutch oven, combine the apple juice and oil. Bring to a boil over medium-high heat. Add the beef. Cook, stirring, for 5 minutes, or until browned. Transfer the beef to a plate.

✺ Add the stock to the pot. Bring to a boil, scraping to loosen any browned bits from the bottom. Add the onions and garlic. Cook, stirring, for 5 minutes, or until the onions are soft but not browned. Add the tomatoes, potatoes, carrots, and beef. Bring to a boil.

✺ Reduce the heat to medium. Cover and cook for 35 minutes, or until the beef is no longer pink in the center. Check by inserting the tip of a sharp knife into 1 piece. Add the salt, black pepper, and red pepper. Stir to combine.

ITALIAN SAUSAGE AND WHITE BEAN STEW

Thrifty cooks love this robust stew because it's a meal in a bowl for just 28¢ a serving. Red potatoes, which are sometimes available frozen, are particularly good in this dish. If you're using cooked navy beans that you've frozen yourself in 1-cup packages, there's no need to thaw before adding to the soup. Use a serrated knife to chop the frozen sausage.

Makes 6 cups

Per 1 cup
Calories 176
Total fat 1.4 g.
Saturated fat 0 g.
Cholesterol 17 mg.
Sodium 445 mg.
Fiber 4.1 g.

Cost per serving

28¢

1 cup frozen chopped onions
4 ounces frozen low-fat hot Italian
    turkey sausage, chopped
1 cup frozen chopped carrots
1 cup frozen diced potatoes
1 cup frozen cooked navy beans (page 177)
4 large cloves garlic, minced
¼ cup all-purpose flour
3 cups frozen defatted Chicken Stock
    (page 61), thawed
½ teaspoon dried thyme
½ teaspoon ground black pepper
¼ teaspoon salt
½ cup nonfat sour cream

※ Coat a Dutch oven with no-stick spray and place over medium-high heat until hot. Add the onions and sausage. Cook, stirring, for 5 minutes, or until the onions are soft and the sausage is browned. Add the carrots, potatoes, beans, garlic, flour, and ½ cup of the stock. Cook, stirring, for 2 minutes. Add the thyme and the remaining 2½ cups stock. Bring to a boil.

※ Cook for 20 minutes, or until the stew is thick and the sausage is no longer pink in the center. Check by inserting the tip of a sharp knife into 1 piece. Add the pepper and salt. Stir to combine. Top each serving with the sour cream.

Per 1 cup
Calories 205
Total fat 1.9 g.
Saturated fat 0.5 g.
Cholesterol 17 mg.
Sodium 296 mg.
Fiber 4.6 g.

Cost per serving

36¢

KITCHEN TIP

To freeze, pack the
cooled soup in a
freezer-quality
plastic container.
To use, thaw
overnight in the
refrigerator.
Transfer to a
saucepan. Cover
and cook, stirring
frequently, over
low heat for
15 minutes, or
until hot.

LENTIL-RICE STEW WITH TURKEY SAUSAGE

Turkey sausage is often a lean alternative to pork sausage (check the label to make sure), and it gives surprisingly good flavor. This thick stew features lentils and vegetables in a curry broth. With rye bread and a green salad, it makes a great meal at less than $1 a serving.

8   ounces low-fat Italian turkey sausage
2   cups chopped green cabbage
1   cup chopped onions
1   green pepper, chopped
3   cloves garlic, minced
4   cups frozen defatted Chicken Stock
     (page 61), thawed
1   cup dried lentils
½   cup rice
1   teaspoon curry powder
¼   teaspoon ground black pepper
2   tablespoons minced fresh parsley

❋ Coat a Dutch oven with no-stick spray and place over medium-high heat until hot. Add the sausage. Cook, stirring, for 5 minutes, or until browned. Add the cabbage, onions, green peppers, garlic, and 1 cup of the stock. Cook, stirring, for 5 minutes, or until the onions are soft but not browned.

❋ Add the lentils, rice, curry powder, and the remaining 3 cups stock. Bring to a boil. Reduce the heat to medium. Cover and cook, stirring occasionally, for 25 minutes, or until the lentils are soft and the stew is thick. Add the black pepper and parsley. Stir to combine.

BAG BARGAINS WITH FREEZER SOUPS

Always allow soup to cool thoroughly before freezing. You can set the hot soup pot in a sink full of very cold water to accelerate the cooling. Then refrigerate for a few hours to chill completely. This prevents the soup's intense heat from damaging other items in either the refrigerator or freezer. It also minimizes the opportunity for bacteria to grow in the center of a large pot of slowly cooling broth.

Scout for sturdy freezer-quality containers for your soups at discount stores. The handiest sizes are 1-cup and 2-cup containers.

Line each container with a freezer-quality plastic bag, pressing the bag tightly against the sides of the container. Add the cooled soup, leaving 1 inch of head space for expansion. Freeze solid, then remove the bag. Clearly label and date each bag. This allows you to immediately reuse the container for another recipe.

To thaw, run hot water over the outside of the bag to loosen the frozen soup. Transfer to a medium saucepan and reheat over low heat, adding a little water, if needed, to prevent scorching. Stir the soup often.

If a cream or pureed soup separates on thawing, just process it briefly in a blender or food processor until smooth.

Soups, Stews, and Chowders

Per 1 cup
Calories 125
Total fat 2 g.
Saturated fat 0.2 g.
Cholesterol 35 mg.
Sodium 145 mg.
Fiber 3.1 g.

Cost per serving

87¢

KITCHEN TIP

Get full value from any bunch of herbs you purchase. After using the amount you need for a recipe, chop and freeze the remaining leaves in a small freezer-quality plastic bag. Cilantro and other herbs will remain green and flavorful. Use the herbs while still frozen.

SOUTHWESTERN FISH STEW

With the major ingredients coming from your freezer, this can be a hearty spur-of-the-moment dinner in 25 minutes—for less than $1 a serving. Not only is it more convenient to use fish fillets directly from the freezer, it's actually easier to cut the fillets into cubes when it's frozen. The fish stock can be thawed in just minutes in the microwave.

1½ cups frozen chopped onions
 3 cloves garlic, minced
 2 teaspoons oil
 1 can (28 ounces) reduced-sodium whole
    tomatoes, chopped (with juice)
 2 cups frozen diced potatoes
 2 cups frozen defatted Fish Stock
    (page 61), thawed
 1 cup frozen sliced carrots
 8 ounces frozen cod or haddock fillets,
    cut into 1" cubes
 1 small jalapeño pepper, seeded and minced
    (wear plastic gloves when handling)
 1 bay leaf
 4 ounces frozen uncooked peeled medium shrimp,
    chopped
⅓ cup chopped fresh cilantro
½ teaspoon grated orange rind
½ teaspoon ground black pepper
¼ teaspoon salt

❋ In a Dutch oven, combine the onions, garlic, and oil. Cook, stirring, over medium-high heat for 5 minutes, or until the onions are soft but not browned. Add the tomatoes, potatoes, stock, carrots, cod or haddock, jalapeño peppers, and bay leaf. Cover and cook, stirring occasionally, for 15 minutes. Remove and discard the bay leaf.

❋ Add the shrimp, cilantro, orange rind, black pepper, and salt. Cook, stirring frequently, for 5 minutes, or until the shrimp are bright pink.

Corn Chowder with Green Chili Peppers

This Southwestern version of corn chowder is slightly spicy and very hearty. Look for corn on sale in season for as low as a dozen ears for $1.99. You can scrape the kernels off the cobs, store in freezer-quality plastic bags, and freeze for up to 1 year.

1 cup frozen chopped sweet red or green peppers
1 cup frozen chopped onions
1 cup frozen cubed raw potatoes
½ cup frozen sliced carrots
½ cup thinly sliced celery
2 tablespoons canned diced green chili peppers
1 tablespoon minced garlic
1 teaspoon olive oil
2 tablespoons all-purpose flour
½ teaspoon ground cumin
6 cups frozen defatted Chicken Stock (page 61), thawed
2 cups frozen whole kernel corn
1 cup evaporated skim milk
3 tablespoons minced fresh parsley or cilantro
¼ teaspoon ground black pepper

✳ In a Dutch oven, combine the red or green peppers, onions, potatoes, carrots, celery, chili peppers, garlic, and oil. Cook, stirring, over medium-high heat for 5 minutes, or until the onions are soft but not browned. Add the flour and cumin. Cook, stirring, for 2 minutes. Add the stock and corn. Bring to a boil.

✳ Reduce the heat to medium. Cover and cook for 10 minutes, or until the potatoes are tender. Check by inserting the tip of a sharp knife into 1 cube. Add the milk, parsley or cilantro, and black pepper. Stir to combine. Bring just to a boil.

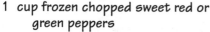

Makes 6 cups

Per 1 cup
Calories 155
Total fat 1.7 g.
Saturated fat 0.3 g.
Cholesterol 2 mg.
Sodium 81 mg.
Fiber 3.2 g.

Cost per serving

62¢

Soups, Stews, and Chowders

CREATIVE CROUTONS

Stash stale or leftover bread in the freezer to use for croutons. These recipes make about 2 cups and save you $2 or more over store-bought boxed croutons.

❋ *Garlic-Herb Croutons.* Preheat the oven to 300°F. Cube 4 slices of bread. Spread on a baking sheet. Coat lightly with no-stick spray. Toss and coat lightly with no-stick spray. Sprinkle with 1 teaspoon minced garlic, ¼ teaspoon dried oregano, and ⅛ teaspoon salt. Bake for 20 minutes, or until crisp. Store in an airtight container.

❋ *Italian Pepper Croutons.* Preheat the oven to 300°F. Cube 4 slices of bread. Spread on a baking sheet. Coat lightly with no-stick spray. Toss and coat lightly with no-stick spray. Sprinkle with 1 teaspoon dried basil, ½ teaspoon dried oregano, ½ teaspoon ground black pepper, and ⅛ teaspoon salt. Bake for 20 minutes, or until crisp. Store in an airtight container.

❋ *Parmesan Croutons.* Preheat the oven to 300°F. Cube 4 slices of bread. Spread on a baking sheet. Coat lightly with no-stick spray. Turn over with a spatula and again coat lightly with no-stick spray. Sprinkle with ¼ cup grated Parmesan cheese, 1 teaspoon olive oil, and ⅛ teaspoon salt. Bake for 20 minutes, or until crisp. Store in an airtight

CURRIED SQUASH-APPLE CHOWDER

Curry powder and ginger give plenty of snap to this creamy soup. Halve the squash and microwave it for 2 minutes to make it easier to peel with a sharp kitchen knife.

1   small sweet red pepper, chopped
1   medium onion, chopped
¼   cup apple juice
2   tablespoons minced garlic
3   cups frozen defatted Chicken Stock
      (page 61), thawed
3   cups peeled and diced butternut squash
2   tart green apples, peeled and chopped
1   cup sliced carrots
1   cup diced potatoes
1   teaspoon curry powder
1   teaspoon grated fresh ginger
½   teaspoon ground red pepper
½   cup nonfat plain yogurt
⅓   cup minced fresh parsley

⁂ In a Dutch oven, combine the chopped peppers, onions, apple juice, and garlic. Cook, stirring, over medium-high heat for 5 minutes, or until the onions are soft but not browned. Add the stock, squash, apples, carrots, potatoes, curry powder, ginger, and ground red pepper. Bring to a boil.

⁂ Reduce the heat to medium. Cover and cook for 20 minutes, or until the squash is very soft. Check by inserting the tip of a sharp knife into 1 piece.

⁂ Process in a blender or food processor until smooth. Return to the pot and heat through. Top each serving with a dollop of the yogurt. Sprinkle with the parsley.

Makes 6 cups

Per 1 cup
Calories 135
Total fat 0.5 g.
Saturated fat 0.1 g.
Cholesterol 0 mg.
Sodium 43 mg.
Fiber 5.5 g.

Cost per serving

41¢

KITCHEN TIP

To freeze, pack cooled soup in a freezer-quality plastic container. To use, thaw overnight in the refrigerator. If it separates after thawing, process briefly in a blender or food processor. Transfer to a saucepan. Cover and cook, stirring frequently, over low heat for 15 minutes, or until hot.

To freeze, pack the
cooled stock in
1-cup or larger
freezer-quality
plastic containers.
Or pour the cooled
stock into ice-cube
trays, freeze until
solid, then
transfer the cubes
to a freezer-
quality plastic bag.
To use, thaw in the
microwave or in a
covered saucepan
set over medium
heat.

Makes 12 cups

Per 1 cup
Calories 2
Total fat 0 g.
Saturated fat 0 g.
Cholesterol 0 mg.
Sodium 2 mg.
Fiber 0.1 g.

Cost per serving

3¢

HOMEMADE STOCKS FOR SOUPS AND STEWS

Making your own stock not only helps you save money but also puts you in control of the sodium and fat content of your soups.

Thrift experts encourage saving any leftover bones or scraps from meat and fish to make great stock. Many butchers also give away beef or lamb bones and other great stock ingredients. Look for free fish heads or fish bones at your fish market. Or save your own shrimp shells for flavorful stock; use in addition to or in place of the fish bones.

Cook stock in an uncovered Dutch oven to allow some of the water to evaporate and the flavors to concentrate.

Each of the following recipes will provide you with about 12 cups of stock. You can save time by cooking a double batch and freezing the extra stock in 1-cup containers for later use.

VEGETABLE STOCK

14 cups water
4 stalks celery (with leaves), chopped
4 carrots, chopped
1 cup chopped green cabbage
1 cup mushrooms
1 onion, chopped
12 cloves garlic
10 sprigs parsley
2 bay leaves

In a Dutch oven, combine the water, celery, carrots, cabbage, mushrooms, onions, garlic, parsley, and bay leaves; bring to a boil over medium-high heat. Reduce the heat to medium; cook for 1 hour, or until the stock is golden brown. Strain the stock through a colander, pressing the vegetables lightly to extract their flavor. Discard the vegetables.

CHICKEN STOCK

14 cups water
4 pounds chicken pieces (thighs, drumsticks, backs, and necks)
2 large carrots, quartered
2 small onions, unpeeled
6 cloves garlic
2 bay leaves

In a Dutch oven, combine the water, chicken, carrots, onions, garlic, and bay leaves; bring to a boil over medium-high heat. Skim the foam from the top; reduce the heat to low. Cook for 2 hours, or until the stock has a rich chicken flavor.

Strain the stock through a colander, pressing the ingredients to extract their flavor. Save the chicken for another use; discard the vegetables. Refrigerate the stock overnight; defat by skimming off and discarding any solidified fat before using.

Makes 12 cups

Per 1 cup
Calories 14
Total fat 0.4 g.
Saturated fat 0.1 g.
Cholesterol 6 mg.
Sodium 6 mg.
Fiber 0.1 g.

Cost per serving

13¢

FISH STOCK

14 cups water
2 pounds fish heads or fish bones
2 large carrots, quartered
2 small onions, unpeeled
6 cloves garlic
2 bay leaves

In a Dutch oven, combine the water, fish heads or bones, carrots, onions, garlic, and bay leaves; bring to a boil over medium-high heat. Skim the foam from the top; reduce the heat to low. Cook for 2 hours, or until the stock has a rich fish flavor.

Strain the stock through a colander, pressing the ingredients to extract their flavor. Discard the solids. Refrigerate the stock overnight; defat by skimming off any solidified fat before using.

Makes 12 cups

Per 1 cup
Calories 10
Total fat 0.3 g.
Saturated fat 0 g.
Cholesterol 0 mg.
Sodium 0 mg.
Fiber 0 g.

Cost per serving

2¢

NEW ENGLAND FISH CHOWDER

We borrowed a thrifty trick from Maine cooks to intensify the flavor of chowder: We simmer the corncobs along with the vegetables. When fresh corn is out of season, you can still enjoy this chowder made with 3 cups frozen whole kernel corn.

1   medium onion, chopped
2   cloves garlic, minced
3   tablespoons shredded carrots
2   tablespoons minced celery
1   teaspoon olive oil
2   tablespoons all-purpose flour
2   ears corn
2   cups frozen defatted Vegetable Stock (page 60)
      or Fish Stock (page 61), thawed
2   cups diced potatoes
2   cups skim milk
1   pound cod fillets, cut into 1" pieces
¼   teaspoon salt
¼   teaspoon ground black pepper

⁂ In a Dutch oven, combine the onions, garlic, carrots, celery, and oil. Cook, stirring, over medium-high heat for 5 minutes, or until the onions are soft but not browned. Add the flour. Cook, stirring, for 2 minutes.

⁂ With a sharp knife, cut the kernels from the corncobs. Add both the kernels and the corncobs to the pot. Add the stock and potatoes. Bring to a boil.

⁂ Reduce the heat to medium. Cover and cook for 20 minutes, or until the potatoes are tender. Check by inserting the tip of a sharp knife into 1 piece.

⁂ Remove and discard the corncobs. Add the milk, cod, salt, and pepper. Bring to a boil. Cook for 10 minutes, or until the chowder is thick and the fish flakes easily when tested with a fork.

Per 1 cup
Calories 199
Total fat 2 g.
Saturated fat 0.4 g.
Cholesterol 37 mg.
Sodium 183 mg.
Fiber 1.9 g.

Cost per serving

64¢

KITCHEN TIP

To freeze, pack the cooled soup in a freezer-quality plastic container. To use, thaw overnight in the refrigerator. Transfer to a saucepan. Cover and cook, stirring frequently, over low heat for 15 minutes, or until hot.

WISCONSIN CHEESE CHOWDER

Cheese-lovers give four stars to this creamy chowder, which is thick with extra-sharp Cheddar and Swiss. Pureeing a small amount of the soup gives it a creamier texture.

1 small sweet red or green pepper, minced
1 small onion, minced
1 cup sliced celery
¼ cup apple juice
1 teaspoon minced garlic
⅓ cup all-purpose flour
2 cups frozen defatted Chicken Stock
      (page 61), thawed
4 cups diced potatoes
1 cup skim milk
¾ cup shredded extra-sharp low-fat Cheddar cheese
¼ cup shredded nonfat Swiss cheese

✸ In a Dutch oven, combine the peppers, onions, celery, apple juice, and garlic. Cook, stirring, over medium-high heat for 5 minutes. Add the flour and ½ cup of the stock. Cook, stirring, for 2 minutes.

✸ Add the potatoes and the remaining 1½ cups stock. Bring to a boil. Reduce the heat to medium. Cover and cook for 20 minutes, or until the potatoes are tender. Check by inserting the tip of a sharp knife into 1 piece.

✸ Transfer 1 cup of the soup to a blender or food processor. Process until smooth. Return to the pot. Add the milk, Cheddar, and Swiss. Stir to combine. Cook, stirring, for 3 minutes, or until the cheese melts.

Makes 9 cups

Per 1 cup
Calories 124
Total fat 1.5 g.
Saturated fat 1.1 g.
Cholesterol 6 mg.
Sodium 132 mg.
Fiber 1.4 g.

Cost per serving

19¢

KITCHEN TIP

To freeze, pack the cooled soup in a freezer-quality plastic container. To use, thaw overnight in the refrigerator. If it separates after thawing, process in a blender or food processor until smooth. Transfer to a saucepan. Cover and cook, stirring frequently, over low heat for 4 to 5 minutes, or until hot. Be careful not to scorch.

Soups, Stews, and Chowders

CHAPTER 4

CHICKEN AND TURKEY

....

Poultry comes home to roost with classic flavor and incredible versatility for as little as 29 cents a pound. And since chicken and turkey freeze better and longer than most red meats or fish—as long as 12 months for whole chickens and turkeys—you can scout out sales and buy in quantity whenever the price drops.

Many cooks believe that poultry holds its flavor and texture best if it's frozen on the bone with the skin on because poultry skin acts as an insulator against moisture loss. You can, however, skin and bone pieces, especially chicken and turkey breast, for quick-thawing individual portions.

Leftover cooked poultry can be frozen on a baking sheet then packaged in small freezer-quality bags for up to six weeks. This method gives you loose-packed pieces to add to soups, stews, and stir-fries.

Good taste, variety, and value—you get them all with the flock of recipes in this chapter.

CHICKEN SAUSAGE SKILLET SUPPER

Makes 6 servings

Per serving
Calories 326
Total fat 13.2 g.
Saturated fat 4.3 g.
Cholesterol 32 mg.
Sodium 503 mg.
Fiber 1.4 g.

Cost per serving

50¢

Most supermarkets sell a variety of lean sausages—including those made with chicken and turkey—that are economically priced. Stock up for the freezer when they're on sale. Use a serrated knife to cut the frozen sausage into small pieces, and it will thaw during cooking. Serve with hot rice and a green salad.

1 teaspoon olive oil
1 pound frozen low-fat reduced-sodium
    chicken sausage, chopped
¾ cup frozen defatted Chicken Stock
    (page 61), thawed
2 cups frozen chopped onions
1 teaspoon minced garlic
2 tablespoons balsamic vinegar
¾ cup frozen chopped green peppers
1 cup long-grain white rice

✳ In a 10″ no-stick skillet over medium-high heat, combine the oil and sausage. Cook for 20 minutes, turning frequently, or until golden brown. Place the sausage on a plate; cover to keep warm.

✳ Add the stock to the skillet. Bring to a boil, scraping to loosen any browned bits from the bottom. Reduce the heat to medium. Add the onions and garlic. Cook, stirring occasionally, for 15 minutes, or until the onions are soft but not browned. Add the vinegar and peppers. Increase the heat to medium-high. Cook for 5 to 6 minutes, or until the peppers are tender.

✳ Meanwhile, cook the rice according to the package directions.

✳ Add the sausage to the skillet. Cover and cook for 5 minutes, or until hot. Add the rice and stir well to combine.

CALIFORNIA PIZZA

Skinless, boneless chicken breasts and frozen vegetables make this pizza a snap to put together. At only 37¢ a serving, it saves you plenty over takeout prices.

1 cup warm water (about 110°F)
1 tablespoon or 1 package active dry yeast
1 tablespoon sugar
½ cup yellow cornmeal
¼ teaspoon salt
2 tablespoons oil
2½ cups all-purpose flour
6 ounces frozen skinned and boned chicken breasts
½ cup frozen chopped sweet red peppers
½ cup frozen chopped onions
½ cup chopped fresh basil
2 cloves garlic, minced
½ cup shredded low-fat Monterey Jack cheese
½ cup grated Parmesan cheese

❋ In a large bowl, combine the water, yeast, and sugar; stir well. Let stand in a warm place for 5 minutes, or until foamy. Stir in the cornmeal, salt, and 1 tablespoon of the oil. Stir in up to 2¼ cups of the flour to make a kneadable dough.

❋ Turn the dough out onto a lightly floured surface. Knead, adding more of the remaining ¼ cup flour as necessary, for 10 minutes, or until smooth and elastic. Coat another large bowl with no-stick spray. Add the dough and turn to coat all sides. Cover loosely with a kitchen towel and set in a warm place for 15 minutes.

❋ Preheat the oven to 450°F. Coat 2 baking sheets with no-stick spray. Divide the dough in half; roll each half into a 12" circle. Place the circles on the sheets. Brush the surface of each pizza with the remaining 1 tablespoon oil.

❋ Thinly slice the chicken and place on the pizza. Top with the peppers, onions, basil, and garlic. Sprinkle with the Monterey Jack and Parmesan.

❋ Bake for 20 minutes, or until the crusts are golden brown and the chicken is no longer pink in the center. Check by inserting the tip of a sharp knife into 1 piece of chicken.

Per serving
Calories 277
Total fat 6.5 g.
Saturated fat 1.7 g.
Cholesterol 17 mg.
Sodium 256 mg.
Fiber 2.5 g.

Cost per serving

37¢

Chicken and Turkey

Per patty
Calories 79
Total fat 2 g.
Saturated fat 0.5 g.
Cholesterol 32 mg.
Sodium 111 mg.
Fiber 0.4 g.

Cost per serving

20¢

KITCHEN TIP

To freeze, place the cooled cooked patties on a tray. Put in the freezer for several hours, or until solid. Transfer to a freezer-quality plastic bag or container. Pack the sauce in a separate freezer-quality plastic container. To use, thaw both overnight in the refrigerator. Reheat the patties in a covered 10" no-stick skillet over medium heat, turning once, for 5 minutes, or until hot. Reheat the sauce in a covered medium saucepan, stirring frequently, for 5 minutes, or until bubbling.

HERBED CHICKEN PATTIES

This Southern-style entrée is embarrassingly easy to prepare. The patties can be made in quantity to freeze. Cornbread adds an authentic note (and is a good use for leftovers), but you can also use soft bread crumbs to bind the patties.

1   pound ground chicken breast
½   cup crumbled cornbread or soft bread crumbs
½   cup chopped scallions
¼   cup chopped fresh parsley
2   cloves garlic, minced
1   teaspoon dried thyme
½   teaspoon hot-pepper sauce
½   teaspoon dried sage
½   teaspoon ground black pepper
¼   teaspoon salt
1   teaspoon oil
2   cups frozen defatted Chicken Stock (page 61), thawed

🌸 In a medium bowl, combine the chicken, cornbread or bread crumbs, scallions, parsley, garlic, thyme, hot-pepper sauce, sage, black pepper, and salt. Mix well and form into 8 thick patties. Place on a plate; cover and refrigerate for 30 minutes.

🌸 Coat a 10" no-stick skillet with no-stick spray and place over medium-high heat. Add ½ teaspoon of the oil. When the oil is hot, add 4 of the patties. Cook for 6 minutes, or until golden brown. Turn and cook for 6 minutes, or until the patties are golden brown and no longer pink in the center. Check by inserting the tip of a sharp knife into 1 patty. Transfer to a plate. Cook the remaining 4 patties in the remaining ½ teaspoon oil. Transfer to a plate.

🌸 Add the stock to the skillet, scraping to loosen any browned bits from the bottom. Bring to a boil over high heat. Cook for 5 minutes, or until the stock is reduced by half. Spoon over the patties.

CHICKEN LO MEIN

Instead of ordering takeout at your local Chinese restaurant, whip up this easy noodle dish and save $3 a serving. If you make a double recipe, the extra lo mein keeps for up to 3 months in the freezer.

4   ounces spaghetti or fettuccine
½   cup frozen defatted Chicken Stock
      (page 61), thawed
¼   cup reduced-sodium soy sauce
2   tablespoons packed brown sugar
2   tablespoons frozen orange juice concentrate,
      thawed
3   cloves garlic, minced
1   tablespoon grated fresh ginger
1   sweet red pepper, diced
1   small onion, diced
8   ounces chicken breasts, skinned, boned, and cut
      into strips
½   cup snow peas, cut in thirds diagonally
2   scallions, thinly sliced

- Cook the spaghetti or fettuccine in a large pot of boiling water according to the package directions. Drain well.

- In a medium bowl, combine the stock, soy sauce, brown sugar, and orange juice concentrate. Mix well.

- Coat a 10″ no-stick skillet with no-stick spray and place over medium-high heat until hot. Add the garlic and ginger. Cook, stirring, for 2 minutes, or until fragrant. Add the peppers and onions. Cook, stirring, for 5 minutes, or until the onions are soft but not browned.

- Add the chicken. Cook, stirring, for 3 to 5 minutes, or until the chicken is no longer pink in the center. Check by inserting the tip of a sharp knife into 1 strip.

- Add the snow peas, pasta, and the stock mixture. Cook, stirring occasionally, for 1 minute, or until the sauce thickens. Sprinkle with the scallions.

Makes 4 servings

Per serving
Calories 237
Total fat 1.7 g.
Saturated fat 0.5 g.
Cholesterol 23 mg.
Sodium 627 mg.
Fiber 2.9 g.

Cost per serving

99¢

KITCHEN TIP

To freeze, pack the cooled lo mein in a freezer-quality plastic container. To use, thaw overnight in the refrigerator. Place in a 10″ no-stick skillet. Cover and cook for 15 minutes, or until hot.

CHICKEN AND VEGETABLE COUSCOUS

Makes 4 servings

Per serving
Calories 399
Total fat 2 g.
Saturated fat 0.7 g.
Cholesterol 45 mg.
Sodium 230 mg.
Fiber 6.4 g.

Cost per serving

$1.01

Keep skinned and boned chicken breasts (see "Bargains to Crow About") in the freezer for quick meals like this. You don't have to thaw the chicken. Just slice it with a serrated knife and cook. Serve it with warmed pita bread, a tossed salad, and a side dish of nonfat plain yogurt spiced with ground cumin.

2	tablespoons dry bread crumbs
½	teaspoon ground black pepper
¼	teaspoon salt
¼	teaspoon crushed dried rosemary
10	ounces frozen skinned and boned chicken breasts, cut into ½" strips
½	cup frozen defatted Chicken Stock (page 61), thawed
1	cup frozen chopped onions
1	cup frozen sliced carrots
1	cup peeled and diced sweet potatoes
4	cloves garlic, minced
1	can (8 ounces) reduced-sodium whole tomatoes, chopped (with juice)
1	cup couscous

✻ In a shallow bowl, combine the bread crumbs, pepper, salt, and rosemary. Dredge the chicken in the mixture, coating well.

✻ Coat a 10″ no-stick skillet with no-stick spray and place over medium-high heat until hot. Add the chicken. Cook, stirring frequently, for 5 minutes, or until the chicken is no longer pink in the center. Check by inserting the tip of a sharp knife into 1 strip. Place the chicken on a plate; cover to keep warm.

✻ Add the stock to the skillet. Bring to a boil, scraping to loosen any browned bits from the bottom. Add the onions, carrots, sweet potatoes, and garlic. Cook, stirring, for 5 minutes, or until the onions are soft but not browned. Add the tomatoes (with juice) and chicken. Cover and cook for 10 minutes.

✻ Meanwhile, cook the couscous according to the package directions. Serve the chicken and vegetables over the couscous.

BARGAINS TO CROW ABOUT

With a little practice, it's easy to cut up whole chickens that you can buy on sale for as little as 29¢ a pound. Properly packaged chicken parts can be frozen successfully for up to ten months.

Remove the giblets and set aside. Wash the chicken inside and out with cold water to remove any surface bacteria. Pat dry with paper towels. With a sharp knife, cut off the legs, thighs, and wings at the joints.

To keep breast halves on the bone, use a heavy knife or kitchen shears to split the chicken lengthwise down the breast bone. Cut away the back portion and set aside.

For boneless breast halves, hold the knife as close as possible to the rib bones and scrape away the breast meat, lifting it off in one piece with your other hand. Repeat with the other breast half. Remove the skin and discard, if desired. Reserve the bones for stock. Boned and skinned chicken breasts are convenient to have on hand in the freezer. For many recipes, it's not even necessary to thaw them in advance.

Flash-freeze boneless chicken breast halves in convenient individual portions. Place the breast halves in a single layer on a tray. Place in the freezer for 1 to 2 hours, or until almost frozen. Layer the breast halves between sheets of wax paper then place in a freezer-quality plastic bag, pressing out as much air as possible. Return to the freezer. At cooking time, remove the number you need.

Store the chicken parts in a freezer-quality plastic bag, pressing out as much air as possible. Date the outside of the bag. If cutting up several chickens, it's convenient to package like parts together. You can put all the drumsticks in one bag, for instance, so they're recipe-ready.

Remove the livers from the giblets and place in a freezer-quality plastic bag to save for other recipes or for pet food. Place the remaining giblets, with the reserved backs and breast bones, in a large freezer-quality plastic bag. Freeze to use in chicken stock.

CHICKEN PAPRIKA

With individual servings of skinned and boned chicken breasts in your freezer (page 71), you'll save big on preparation time and money when making this streamlined Hungarian dish. You don't have to thaw the chicken. Just use a serrated knife to cut the frozen breasts into strips. Serve over broad noodles.

1½ pounds frozen chicken breasts, skinned, boned, and cut into ½" strips
¼ cup white wine or water
1 cup frozen chopped onions
1 cup frozen chopped sweet red peppers
2 teaspoons minced garlic
1 teaspoon all-purpose flour
½ cup frozen defatted Chicken Stock (page 61), thawed
2 teaspoons paprika
½ teaspoon ground black pepper
¼ teaspoon salt
½ cup low-fat plain yogurt
1 teaspoon cornstarch

❋ Coat a 10" no-stick skillet with no-stick spray and place over medium-high heat until hot. Add the chicken. Cook, stirring frequently, for 5 minutes, or until the chicken is no longer pink in the center. Check by inserting the tip of a sharp knife into 1 strip. Place the chicken on a plate; cover to keep warm.

❋ Add the wine or water to the skillet. Bring to a boil, scraping to loosen any browned bits from the bottom. Add the onions, chopped peppers, and garlic. Cook, stirring frequently, for 5 to 8 minutes, or until the onions are soft but not browned.

❋ Add the flour. Cook, stirring constantly, for 2 minutes. Add the stock, paprika, black pepper, and salt. Cook for 5 minutes, or until the sauce thickens. Add the chicken and cook for 1 to 2 minutes, or until warm.

❋ In a small bowl, combine the yogurt and cornstarch. Mix well. Remove the skillet from the heat. Stir in the yogurt mixture.

PITAS WITH CHICKEN AND TANGY HERB SAUCE

These irresistible sandwiches make an almost effortless meal that you can serve with a robust soup from the freezer. Stockpile individual servings of skinned and boned chicken breasts in your freezer (page 71) for savings and convenience. You don't have to thaw the chicken. Just cube the frozen breasts with a serrated knife and toss in the marinade.

1   cup nonfat plain yogurt
1   tablespoon chopped fresh parsley or
      cilantro
1   teaspoon honey mustard
10 ounces frozen skinned and boned chicken
      breasts, cut into 1" pieces
¼ cup frozen orange juice, concentrate
1   tablespoon curry powder
2   cloves garlic, minced
4   frozen whole-wheat pita bread rounds (6" diameter)
1   small tomato, chopped
1   cup chopped lettuce

※ In a small bowl, combine the yogurt, parsley or cilantro, and mustard. Mix well. Cover and refrigerate for 1 hour, stirring occasionally.

※ In a shallow nonmetal dish, combine the chicken, orange juice concentrate, curry powder, and garlic. Cover and refrigerate for 1 hour, stirring occasionally.

※ Preheat the broiler. Coat a broiler pan with no-stick spray. Remove the chicken from the marinade and reserve the marinade. Place the chicken on the broiler pan. Broil 4" from the heat, basting frequently with the marinade, for 5 minutes, or until the chicken is no longer pink in the center. Check by inserting the tip of a sharp knife into 1 piece. Discard any remaining marinade.

※ Place the pita bread rounds in a toaster oven for 1 minute, or until warm. Cut the pitas in half. Divide the chicken evenly among the pita bread halves. Top with the tomatoes and lettuce. Drizzle with the herb sauce.

Makes 4 servings

Per serving
Calories 298
Total fat 4.3 g.
Saturated fat 1 g.
Cholesterol 64 mg.
Sodium 353 mg.
Fiber 4.2 g.

Cost per serving

94¢

KITCHEN TIP

Get full value from any bunch of herbs you purchase. After using the amount you need for a recipe, chop and freeze the remaining leaves in a small freezer-quality plastic bag. Parsley, cilantro, and other herbs will remain green and flavorful. Use the herbs while still frozen.

CHICKEN BREASTS WITH MINT SAUCE

Makes 4 servings

Per serving
Calories 201
Total fat 3.7 g.
Saturated fat 0.9 g.
Cholesterol 57 mg.
Sodium 401 mg.
Fiber 0 g.

Cost per serving

90¢

KITCHEN TIP

To freeze, pack the cooled cooked chicken in a freezer-quality plastic container. Pack the sauce in a separate freezer-quality plastic container. To thaw, place both in the refrigerator overnight. Briefly process the sauce in a blender or food processor. Place the chicken and sauce in a 12" no-stick skillet. Cover and cook over low heat, stirring occasionally, for 10 minutes, or until hot.

In this recipe, we create a sauce for chicken breasts by reducing the pan juices and blending in nonfat yogurt and mint jelly. During the summer, you can substitute 1 tablespoon chopped fresh mint leaves for the jelly. Serve with basmati rice and green beans.

Juice of 1 lime
2 tablespoons reduced-sodium soy sauce
4 cloves garlic, minced
2 tablespoons honey or packed brown sugar
1 teaspoon olive oil
½ teaspoon ground coriander
4 chicken breast halves (6 ounces each),
    skinned and boned
1 cup nonfat plain yogurt
1 tablespoon mint jelly

❁ In an 8″ × 8″ glass baking dish, combine the lime juice, soy sauce, garlic, honey or brown sugar, oil, and coriander. Mix well. Add the chicken and turn to coat. Cover and refrigerate for 30 minutes, turning occasionally.

❁ Preheat the oven to 375°F. Bake the chicken in the marinade for 30 minutes, or until no longer pink in the center. Check by inserting the tip of a sharp knife into 1 breast. Place the chicken on a plate.

❁ Pour the pan juices into a medium saucepan. Bring to a boil over medium-high heat. Cook, stirring frequently, for 5 minutes, or until reduced by half. Remove from the heat. Stir in the yogurt and jelly. Serve over the chicken.

Island Chicken with Pineapple Salsa

Serve this fast freezer entrée with a side dish of rice or orzo.

1 can (8 ounces) unsweetened crushed
    pineapple (with juice)
1 tablespoon reduced-sodium soy sauce
1 tablespoon honey
2 cloves garlic, minced
¼ teaspoon crushed red-pepper flakes
4 chicken breast halves (6 ounces each),
    skinned and boned
½ cup diced onions
¼ cup packed brown sugar
2 tablespoons lime juice
1 teaspoon minced jalapeño peppers
    (wear plastic gloves when handling)
1 teaspoon minced fresh cilantro

* Strain the pineapple; reserve the juice. Place the pineapple in a medium bowl. Cover and refrigerate.

* Place the pineapple juice in a shallow nonmetal dish. Add the soy sauce, honey, garlic, and red-pepper flakes. Mix well. Add the chicken and turn to coat all sides. Cover and refrigerate for at least 4 hours or up to 24 hours, turning occasionally.

* Preheat the grill or broiler. Coat the grill rack or broiler pan with no-stick spray.

* Remove the pineapple from the refrigerator. Add the onions, brown sugar, lime juice, jalapeño peppers, and cilantro. Mix well. Let stand at room temperature.

* Remove the chicken from the marinade; reserve the marinade. Grill or broil 4″ from the heat for 5 minutes. Turn and cook for 5 minutes, or until the chicken is no longer pink in the center. Check by inserting the tip of a sharp knife into 1 breast.

* Transfer the marinade to a small saucepan. Bring to boil over medium-high heat. Cook for 5 minutes, or until reduced by half. Pour over the chicken. Top with the salsa.

Makes 4 servings

Per serving
Calories 256
Total fat 3 g.
Saturated fat 0.9 g.
Cholesterol 73 mg.
Sodium 221 mg.
Fiber 0.6 g.

Cost per serving

87¢

Kitchen Tip

To freeze, pack the cooled cooked chicken and sauce in a freezer-quality plastic container. Pack the salsa in a separate freezer-quality plastic container. To use, thaw both overnight in the refrigerator. Cover the chicken and microwave on high power for 3 minutes. Top with the salsa.

Per serving
Calories 395
Total fat 3 g.
Saturated fat 0.6 g.
Cholesterol 45 mg.
Sodium 256 mg.
Fiber 9.8 g.

Cost per serving

$1.09

RAGOÛT OF CHICKEN AND TOMATO

A ragoût is a French stew, cooked gently so that the chicken becomes tender but stays juicy. This recipe does double duty. It can be made with ingredients on hand in your freezer. Or you can make double or triple batches from fresh ingredients to put into the freezer for work-free meals to serve on busy weekdays.

4 frozen chicken breast halves on the bone
      (6 ounces each), thawed and skinned
¼ teaspoon salt
¼ teaspoon ground black pepper
1 cup frozen defatted Chicken Stock
      (page 61), thawed
1 can (28 ounces) reduced-sodium whole
      tomatoes, chopped (with juice)
8 ounces no-yolk noodles
1 cup frozen diced onions
1 cup frozen sliced carrots
3 tablespoons water or red wine
1 cup frozen peas
2 teaspoons minced fresh chives

Coat a Dutch oven with no-stick spray and place over medium-high heat until hot. Add the chicken. Sprinkle with the salt and pepper. Cook for 5 minutes. Turn and cook for 5 minutes, or until the chicken is golden brown.

Add the stock and tomatoes (with juice). Bring to a boil. Cover and cook for 20 minutes, or until the chicken is no longer pink in the center. Check by inserting the tip of a sharp knife into 1 piece. Transfer the chicken to a plate; cover to keep warm.

Cook the noodles in a large pot of boiling water according to the package directions; drain well. Cover to keep warm.

Add the onions, carrots, and water or wine to the Dutch oven. Bring to a boil, scraping to loosen any browned bits from the bottom. Cook for 5 minutes, or until the carrots are crisp-tender. Add the peas and chicken. Cover and cook for 3 minutes, or until hot. Serve over the noodles. Sprinkle with the chives.

STUFFED CHICKEN BREASTS

Slowly cooking the mushrooms for the stuffing coaxes out their sweetness. A golden Parmesan crust is the perfect counterpoint to the moist chicken and stuffing.

½ cup frozen defatted Chicken Stock
        (page 61), thawed
2 cups chopped mushrooms
½ cup frozen chopped onions
½ cup shredded nonfat mozzarella cheese
3 tablespoons frozen chopped sweet red
        peppers, minced
8 tablespoons grated Parmesan cheese
4 frozen boned chicken breast halves (6 ounces each),
        skin on
1 tablespoon minced fresh parsley

❋ Preheat the oven to 375°F. Coat a 13" × 9" baking dish with no-stick spray.

❋ Coat a 10" no-stick skillet with no-stick spray and place over medium heat. Add ¼ cup of the stock and bring to a boil. Add the mushrooms and onions. Cook, stirring, for 10 minutes, or until the liquid has evaporated. Transfer to a medium bowl. Stir in the mozzarella, peppers, and 6 tablespoons of the Parmesan.

❋ Cut a thin slit in the thicker side of each chicken breast to form a cavity. Divide the mushroom mixture among the breasts, packing it into the cavities. Secure with wooden toothpicks. Place in the prepared baking dish. Add the remaining ¼ cup stock.

❋ Cover and bake for 25 minutes, or until the chicken is no longer pink in the center. Check by inserting the tip of a sharp knife into 1 breast.

❋ Remove the chicken from the baking dish. Remove and discard the skin. Place the chicken on a baking sheet. Sprinkle with the remaining 2 tablespoons Parmesan. Broil 4" from the heat for 3 to 5 minutes, or until golden. Serve sprinkled with the parsley.

Makes 4 servings

Per serving
Calories 296
Total fat 8 g.
Saturated fat 3.5 g.
Cholesterol 106 mg.
Sodium 418 mg.
Fiber 0.9 g.

Cost per serving

$1.14

Per serving
Calories 199
Total fat 4.5 g.
Saturated fat 2.3 g.
Cholesterol 73 mg.
Sodium 192 mg.
Fiber 1.6 g.

Cost per serving

97¢

SALSA CHICKEN DINNER

Combine chicken breasts and frozen vegetables and you have a meal in less than 10 minutes. Serve with rice or corn tortillas from your freezer.

4   frozen chicken breast halves (6 ounces each),
       thawed, skinned, and boned
½   cup frozen defatted Chicken Stock
       (page 61), thawed
1   cup frozen minced onions
1   cup frozen sliced carrots
1   large tomato, chopped
1   teaspoon minced garlic
¼   cup salsa
¼   teaspoon ground black pepper
¼   cup grated Parmesan cheese

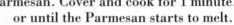

❀ Place each chicken breast between 2 sheets of wax paper. With a meat mallet or the side of a cleaver, pound to ¼″ thickness.

❀ Coat a 10″ no-stick skillet with no-stick spray and place over medium-high heat until hot. Add the chicken. Cook for 2 minutes. Turn the pieces. Add the stock and bring to a boil.

❀ Add the onions, carrots, tomatoes, garlic, salsa, and pepper. Cover and cook for 3 minutes, or until the chicken is no longer pink in the center. Check by inserting the tip of a sharp knife into 1 piece. Sprinkle with the Parmesan. Cover and cook for 1 minute, or until the Parmesan starts to melt.

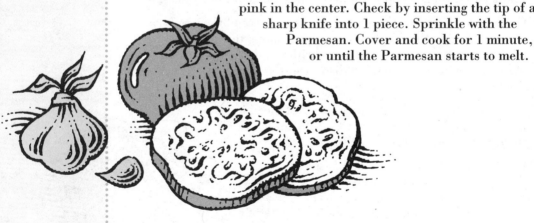

TIMELY TRICKS WITH POULTRY LEFTOVERS

Leftover cooked chicken or turkey freezes well and can be thawed overnight in the refrigerator for one of these fast entrées.

● *Bistro Sandwich.* Split and hollow out a French or Italian hard roll and pack the bottom half with shredded cooked chicken or turkey, chopped red onions, chopped lettuce or spinach, and chopped tomatoes. Sprinkle with balsamic vinegar. Replace the top half of the roll, wrap well, and refrigerate for 30 minutes to blend the flavors.

● *Curried Turkey Omelet.* Fill an egg-white omelet with shredded cooked chicken or turkey, minced scallions, a sprinkle of curry powder, and shredded nonfat Monterey Jack cheese.

● *Hot Wrap.* Roll a flour tortilla around a filling of shredded cooked chicken or turkey, minced scallions, minced green peppers, shredded nonfat mozzarella, and a dash of hot-pepper sauce. Microwave until the mozzarella melts.

● *Poultry Pasta.* Toss hot cooked pasta with a small amount of olive oil. Add shredded cooked chicken or turkey, chopped scallions, and grated Parmesan cheese.

● *Perfect Peking.* Combine shredded cooked chicken or turkey with minced onions, minced celery, chopped water chestnuts, and shredded carrots. Cook in a small amount of defatted chicken stock until the vegetables are tender. Add a dash of hot-pepper sauce and hoisin sauce to taste. Serve over rice.

Per taco
Calories 58
Total fat 4 g.
Saturated fat 0.9 g.
Cholesterol 28 mg.
Sodium 265 mg.
Fiber 0.5 g.

Cost per serving

38¢

CHICKEN AND CHEESE SOFT TACOS

To save fat and calories, these tacos are made with soft rather than deep-fried tortillas. Use chicken thighs for this recipe—at $1.28 per pound, on the bone, you save 70¢ over the same amount of chicken breasts. Just skin and bone them yourself and pack in the freezer for convenient, thrifty main dishes. Take the tortillas out of the freezer when you start preparing the filling, and they'll thaw in minutes.

8 ounces frozen skinned and boned chicken thighs,
    cut into thin strips
½ cup frozen defatted Chicken Stock
    (page 61), thawed
1½ cups frozen diced onions
1 cup frozen chopped green peppers
2 cloves garlic, minced
1 teaspoon minced jalapeño peppers
    (wear plastic gloves when handling)
½ cup medium-hot salsa
2 teaspoons ground cumin
8 frozen corn tortillas (6" diameter)
1 cup shredded low-fat Monterey Jack cheese

✵ Preheat the oven to 400°F. Coat a 13″ × 9″ baking dish with no-stick spray.

✵ Coat a 10″ no-stick skillet with no-stick spray and place over medium-high heat until hot. Add the chicken. Cook, stirring, for 5 minutes, or until the chicken is golden brown and no longer pink in the center. Check by inserting the tip of a sharp knife into 1 strip. Place the chicken on a plate; cover to keep warm.

✵ Add the stock to the skillet. Bring to a boil, scraping to loosen any browned bits from the bottom. Add the onions, green peppers, garlic, and jalapeño peppers. Cook, stirring frequently, for 5 minutes, or until the onions are golden brown.

✵ Add the salsa, cumin, and chicken. Cook, stirring frequently, for 3 minutes, or until the liquid has almost evaporated.

✵ Spoon into the tortillas. Divide ½ cup of the Monterey Jack among the tortillas. Roll up and arrange, seam side down, in the prepared baking dish. Sprinkle with the remaining ½ cup Monterey Jack. Bake for 10 to 12 minutes, or until the cheese melts.

CHICKEN AND VEGETABLE PASTA AU GRATIN

For less than 60¢ a serving, this hearty Midwestern harvest casserole makes a fast weeknight family meal. It can be doubled and the extras frozen for up to 1 month.

8 ounces chicken thighs, skinned, boned, and cut into 1" pieces
1 cup chopped onions
1 cup frozen defatted Chicken Stock (page 61), thawed
2 medium carrots, thinly sliced
2 medium parsnips, thinly sliced
1 tablespoon minced garlic
1 teaspoon Italian herb seasoning
¼ cup all-purpose flour
2 cups skim milk
1 cup shredded nonfat mozzarella cheese
8 ounces penne pasta
¾ cup chopped tomatoes
¼ cup dry bread crumbs
¼ cup grated Parmesan cheese

☀ Preheat the oven to 400°F.

☀ Coat a Dutch oven with no-stick spray and place over medium-high heat until hot. Add the chicken, onions, and ¼ cup of the stock. Cook for 5 minutes, or until the onions are soft but not browned.

☀ Add the carrots, parsnips, garlic, Italian herb seasoning, and the remaining ¾ cup stock. Cover and cook for 5 to 8 minutes, or until the vegetables are slightly tender. Add the flour. Cook, stirring constantly, for 2 minutes.

☀ Gradually add the milk and mozzarella. Cook, stirring frequently, for 4 minutes, or until the sauce thickens.

☀ Cook the penne pasta in a large pot of boiling water according to the package directions. Drain well and add to the Dutch oven. Add the tomatoes. Mix well to combine.

☀ Top with the bread crumbs and Parmesan. Bake for 20 minutes, or until the topping is golden brown.

Makes 6 servings

Per serving
Calories 341
Total fat 4.6 g.
Saturated fat 1 g.
Cholesterol 24 mg.
Sodium 327 mg.
Fiber 4.3 g.

Cost per serving

54¢

KITCHEN TIP

To freeze, pack the cooled cooked chicken and pasta in a freezer-quality plastic container. To use, thaw overnight in the refrigerator. Spoon into a 13" × 9" baking dish, cover, and bake at 400°F for 20 minutes, or until bubbling.

FIVE FAST HERB BLENDS FOR CHICKEN

Spend a minute making these easy herb blends, then store in small containers in a cool cupboard or in the freezer, where they'll keep for 6 months. Dress up ordinary chicken breasts or thighs for only pennies a serving by seasoning with your favorite blend.

1. *Cajun Blend*. Combine equal amounts of ground black pepper, ground red pepper, dried thyme, ground anise seeds, and dried basil.

2. *Greek Blend*. Combine equal amounts of dried thyme, dried oregano, and garlic powder.

3. *Italian Blend*. Combine equal amounts of dried sage, dried savory, and crushed fennel seeds.

4. *Mediterranean Blend*. Combine equal amounts of crushed dried rosemary, dried thyme, dried orange rind, and ground black pepper.

5. *South-of-France Blend*. Combine equal amounts of dried basil, dried oregano, and dried thyme.

Oven-Barbecued Chicken Thighs

Homemade barbecue sauce is half the price and has half the sodium of store-bought. Plus, it tastes better. With frozen chicken thighs, frozen chopped onions, and shelf staples, you can make this sensational barbecue fast.

8 frozen chicken thighs, thawed and skinned
¼ cup reduced-sodium ketchup
1 tablespoon packed brown sugar
1 tablespoon cider vinegar
1 tablespoon frozen chopped onions, minced
1 teaspoon minced garlic
½ teaspoon hot-pepper sauce

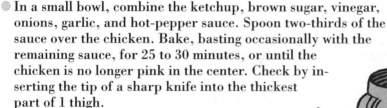

❋ Preheat the oven to 400°F. Cover a large baking sheet with foil. Arrange the chicken on the sheet.

❋ In a small bowl, combine the ketchup, brown sugar, vinegar, onions, garlic, and hot-pepper sauce. Spoon two-thirds of the sauce over the chicken. Bake, basting occasionally with the remaining sauce, for 25 to 30 minutes, or until the chicken is no longer pink in the center. Check by inserting the tip of a sharp knife into the thickest part of 1 thigh.

Makes 4 servings

Per serving
Calories 249
Total fat 11 g.
Saturated fat 3.2 g.
Cholesterol 99 mg.
Sodium 97 mg.
Fiber 0.3 g.

Cost per serving

74¢

Per serving
Calories 300
Total fat 4.6 g.
Saturated fat 1.3 g.
Cholesterol 39 mg.
Sodium 599 mg.
Fiber 2.6 g.

Cost per serving

85¢

KITCHEN TIP

Tortillas keep well
in the freezer and
can even be
refrozen. Often,
you can pry apart
the frozen tortillas
without breaking
them. If not, thaw
the whole package
and remove the
number that you
need. Squeeze the
air out of the bag.
Reseal the bag
tightly and
refreeze.

TURKEY AND GREEN CHILI PEPPER BURRITOS

Replacing beef with turkey in these traditional Southwestern burritos saves 4.5 grams of fat per serving. Remove the tortillas from the freezer before you start making the burrito stuffing. The turkey breast can be quickly thawed in the microwave before cooking.

⅓ cup apple juice
1 cup frozen chopped onions
2 cloves garlic, minced
8 ounces frozen ground turkey breast
½ cup frozen sliced zucchini or yellow
      summer squash
¼ cup frozen diced green or sweet red peppers
¼ cup thinly sliced mushrooms
¼ cup canned diced green chili peppers
¼ cup medium salsa
2 tablespoons minced fresh cilantro
½ teaspoon ground cumin
½ teaspoon ground black pepper
4 frozen flour tortillas (12" diameter)
⅓ cup shredded low-fat Monterey Jack cheese

❋ Preheat the oven to 350°F. Coat an 8" × 8" baking dish with no-stick spray.

❋ Bring the apple juice to a boil in a 10" no-stick skillet over medium-high heat. Add the onions and garlic. Cook, stirring, for 5 minutes, or until the onions are soft but not browned. Add the turkey. Cook, stirring, for 8 minutes, or until the turkey is no longer pink.

❋ Add the zucchini or squash, green or red peppers, mushrooms, chili peppers, and salsa. Cook, stirring, for 3 minutes, or until the mixture is hot. Stir in the cilantro, cumin, and black pepper.

❋ Divide the turkey mixture among the tortillas. Top with the Monterey Jack. Roll up and place, seam side down, in the prepared baking dish. Cover with foil. Bake for 15 minutes, or until hot.

TURKEY SAUSAGE STEW

Frozen vegetables cut fat and cost by replacing most of the meat in this classic French one-pot meal. Chopping the frozen sausage is an easy task with a serrated knife.

12 ounces frozen low-fat turkey sausage, chopped
4 cups frozen defatted Chicken Stock (page 61), thawed
1 cup frozen chopped onions
1 cup frozen sliced carrots
1 cup frozen chopped cauliflower
1 can (16 ounces) reduced-sodium whole tomatoes, chopped (with juice)
2 cups frozen cooked navy beans (page 177)
1 small eggplant, coarsely chopped
1 tablespoon minced garlic
2 teaspoons Italian herb seasoning
½ teaspoon ground cumin
¼ cup grated Parmesan cheese

❋ Coat a Dutch oven with no-stick spray and place over medium-high heat until hot. Add the sausage. Cook, stirring frequently, for 5 to 8 minutes, or until the sausage is golden brown. Add ½ cup of the stock. Bring to a boil, scraping to loosen any browned bits from the bottom.

❋ Add the onions, carrots, and cauliflower. Cook, stirring, for 5 minutes, or until the onions are soft but not browned. Add the tomatoes (with juice), beans, eggplant, garlic, Italian herb seasoning, cumin, and the remaining 3½ cups stock. Bring to a boil.

❋ Reduce the heat to medium. Cook, stirring occasionally, for 30 to 40 minutes, or until thick. Serve sprinkled with the Parmesan.

Makes 6 servings

Per serving
Calories 255
Total fat 4.1 g.
Saturated fat 1.6 g.
Cholesterol 37 mg.
Sodium 592 mg.
Fiber 9.9 g.

Cost per serving

69¢

SPECIAL

99¢ lb.

Per serving
Calories 402
Total fat 1.7 g.
Saturated fat 0.4 g.
Cholesterol 57 mg.
Sodium 65 mg.
Fiber 2.3 g.

Cost per serving

92¢

KITCHEN TIP

To freeze, remove
the cooled cooked
turkey and
vegetables from
the skewers. Place
on a tray and
freeze for several
hours, or until
solid. Pack in a
freezer-quality
plastic bag. To
freeze the rice,
pack in a freezer-
quality plastic
container. To use,
thaw both
overnight in the
refrigerator. Place
the rice on a
microwaveable
dish. Top with the
turkey and
vegetables. Cover
and microwave
on high power for
5 minutes, or
until hot.

CURRIED TURKEY KABOBS OVER RICE

This exciting Indian spice combination really perks up the flavor of turkey. If you make these colorful kabobs to freeze for a future meal, it takes just a moment to mix the yogurt sauce at serving time.

1   pound turkey breast, skinned, boned,
      and cut into 1" cubes
½   cup orange juice
1   tablespoon curry powder
½   teaspoon ground cinnamon
½   teaspoon ground coriander
1   cup long-grain white rice
½   cup frozen defatted Chicken Stock
      (page 61), thawed
1   pound frozen pearl onions
1   green pepper, cut into 8 pieces
1   sweet red pepper, cut into 8 pieces
¼   cup nonfat plain yogurt
1   tablespoon minced scallions
¼   teaspoon ground cumin

In a shallow nonmetal dish, combine the turkey, orange juice, curry powder, cinnamon, and coriander. Mix well to combine. Cover and refrigerate for 1 hour, stirring frequently.

Preheat the broiler or grill. Coat a broiler pan or grill rack with no-stick spray.

Cook the rice according to the package directions. Cover to keep warm.

Bring the stock to a boil in a 10" no-stick skillet over medium-high heat. Add the onions. Cook, stirring, for 2 minutes, or until the onions are thawed.

Thread the turkey, onions, green peppers, and red peppers onto four 6" skewers; reserve the turkey marinade.

Broil or grill the kabobs 4" from the heat, basting with the marinade and turning frequently, for 12 minutes, or until the turkey is no longer pink in the center. Check by inserting the tip of a sharp knife into 1 cube. Discard any remaining marinade.

In a small bowl, combine the yogurt, scallions, and cumin. Mix well to combine. Spoon the rice onto a platter. Top with the kabobs. Serve with the yogurt sauce.

HERBED TURKEY AND VEGETABLE HASH

Makes 6 servings

Per serving
Calories 222
Total fat 2.2 g.
Saturated fat 0.5 g.
Cholesterol 51 mg.
Sodium 165 mg.
Fiber 3.1 g.

Cost per serving

70¢

Turkey replaces corned beef in this hash for a savings of 20¢ a serving. For brunch, serve it with a green salad and a poached fruit dessert. If you have fresh parsley on hand, sprinkle over the dish before serving.

1   teaspoon olive oil
1   pound frozen turkey breast, cut into 1" cubes
¼   cup apple juice
3   cups frozen shredded potatoes
¼   cup frozen chopped onions
2   cups frozen sliced zucchini
1   teaspoon dried sage
¼   teaspoon dried thyme
¼   teaspoon salt
⅛   teaspoon ground red pepper
½   cup nonfat sour cream

❋ Coat a 10" no-stick skillet with no-stick spray and place over medium-high heat. Add the oil and heat until hot. Add the turkey. Cook, stirring, for 3 minutes, or until the turkey is no longer pink in the center. Check by inserting the tip of a sharp knife into 1 cube. Place the turkey on a plate; cover to keep warm.

❋ Add the apple juice to the skillet. Bring to a boil, scraping to loosen any browned bits from the bottom. Add the potatoes, onions, zucchini, sage, and thyme. Cook, stirring, for 8 minutes, or until the onions are soft but not browned. Add the salt, pepper, and turkey. Mix well to combine. Serve topped with the sour cream.

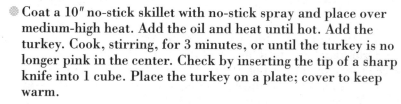

ORANGE-SESAME TURKEY CUTLETS

A teaspoon of dark sesame oil adds extraordinary flavor to these turkey cutlets for only 2¢ per serving. The lean meat is cooked right in its marinade to keep it tender and moist. Serve over rice.

Per serving
Calories 170
Total fat 2.4 g.
Saturated fat 0.5 g.
Cholesterol 77 mg.
Sodium 202 mg.
Fiber 0.2 g.

Cost per serving

65¢

1   cup orange juice
1   tablespoon reduced-sodium soy sauce
1   teaspoon dark sesame oil
1   teaspoon minced garlic
1   pound turkey cutlets
1   teaspoon cornstarch
¾   cup frozen defatted Chicken Stock
     (page 61), thawed
1   tablespoon minced fresh cilantro

*In a shallow nonmetal dish, combine the orange juice, soy sauce, oil, and garlic. Add the turkey. Turn to coat. Cover and refrigerate for 1 hour, turning frequently.

*Coat a 10" no-stick skillet with no-stick spray and place over medium-high heat until hot. Add the turkey in a single layer; reserve the marinade. Cook the turkey for 2 minutes. Turn and cook for 2 minutes, or until golden brown. Place the turkey on a plate; cover to keep warm.

KITCHEN TIP

To freeze, pack the cooled cooked cutlets and sauce in a freezer-quality plastic container. To use, thaw overnight in the refrigerator. Cover and microwave on high power for 5 minutes, or until hot.

*Place the cornstarch in a medium bowl. Add the stock and stir until smooth. Add the reserved marinade and mix well. Reduce the heat to medium. Pour into the skillet. Cook, stirring, for 3 minutes, or until the sauce thickens slightly. Add the turkey. Cook for 5 minutes, or until the turkey is no longer pink in the center. Check by inserting the tip of a sharp knife into the center of 1 cutlet. Serve sprinkled with the cilantro.

BERRIES
PASTA
SALAD
BEANS

TURKEY SCALOPPINE

Thin turkey cutlets cook in less than 10 minutes, making this party dish both fast and economical at less than $1 a serving.

2 tablespoons all-purpose flour
¼ teaspoon salt
¼ teaspoon ground black pepper
1 pound turkey cutlets
1 teaspoon olive oil
¼ cup dry sherry or apple juice
2 cups chopped mushrooms
⅓ cup chopped scallions
1 clove garlic, minced
½ teaspoon grated fresh ginger
¼ cup apricot jam
1 tablespoon cider vinegar
1 teaspoon cornstarch
½ cup frozen defatted Chicken Stock (page 61), thawed

* In a shallow dish, combine the flour, salt, and pepper. Dredge the turkey in the flour mixture, coating each piece well.

* Coat a 10″ no-stick skillet with no-stick spray and place over medium heat. Add the oil and heat until hot. Add the turkey in a single layer. Cook for 3 minutes. Turn and cook for 3 minutes, or until golden brown and no longer pink in the center. Check by inserting the tip of a sharp knife into 1 cutlet. Place the turkey on a plate; cover to keep warm.

* Add the sherry or apple juice to the skillet. Bring to a boil over medium-high heat, scraping to loosen any browned bits from the bottom. Add the mushrooms, scallions, garlic, and ginger. Cook, stirring, for 10 minutes, or until the liquid has evaporated.

* Reduce the heat to medium. Add the jam and vinegar; cook, stirring, for 1 minute.

* Place the cornstarch in a small bowl. Add the stock and stir until smooth. Add to the skillet. Cook, stirring constantly, for 1 minute, or until the sauce is thick. Serve over the turkey.

Per serving
Calories 223
Total fat 2.4 g.
Saturated fat 0.5 g.
Cholesterol 77 mg.
Sodium 195 mg.
Fiber 1.2 g.

Cost per serving

88¢

KITCHEN TIP

To freeze, pack the cooled cooked turkey and sauce in a freezer-quality plastic container. To use, thaw overnight in the refrigerator. Cover and microwave on high power for 5 minutes, or until hot.

FREEZER SAUCES TOP TURKEY

Serve one of these fast freezer sauces to top grilled or broiled turkey cutlets. Each recipe makes about 1 cup of sauce, enough for 4 servings.

Cran-Orange Compote. In a blender or food processor, combine 1 cup thawed frozen cranberries, $\frac{1}{3}$ cup thawed frozen orange juice concentrate, and $\frac{1}{4}$ cup sugar. Puree.

Great Gravy. In a small saucepan, combine 1 cup thawed frozen defatted Chicken Stock (page 61), 2 tablespoons minced garlic, and 2 tablespoons cornstarch. Mix well. Set the saucepan over medium-high heat. Cook, stirring, for 5 to 8 minutes, or until the sauce thickens. Stir in $\frac{1}{4}$ teaspoon salt and $\frac{1}{4}$ teaspoon ground black pepper.

Mushroom Sauce. Coat a 10" no-stick skillet with no-stick spray and place over medium-high heat until hot. Add 1 cup thawed frozen chopped onions. Cook, stirring, for 5 minutes, or until soft but not browned. Add 1 cup diced mushrooms and 1 cup thawed frozen defatted Chicken Stock. Bring to a boil. Reduce the heat to medium. Cook, stirring frequently, for 20 minutes, or until the mushrooms are very soft. In a cup, mix 1 tablespoon frozen apple juice concentrate and 1 teaspoon cornstarch until smooth. Add to the saucepan. Cook, stirring, for 3 minutes, or until the sauce thickens.

GINGER-SOY TURKEY BREAST

If desired, you can mix the turkey with the marinade and freeze it for future use. To use, thaw and proceed with the recipe.

Makes 6 servings

Per serving
Calories 139
Total fat 1.1 g.
Saturated fat 0.4 g.
Cholesterol 76 mg.
Sodium 190 mg.
Fiber 0.3 g.

Cost per serving

60¢

2 tablespoons cider vinegar
1 tablespoon reduced-sodium soy sauce
1 tablespoon grated fresh ginger
1 tablespoon packed brown sugar
1 tablespoon coarse-grained mustard
1 small turkey breast (2 pounds),
   skinned and boned

❋ In a shallow nonmetal dish, combine the vinegar, soy sauce, ginger, brown sugar, and mustard. Mix well. Add the turkey and turn to coat all sides. Cover and refrigerate for at least 1 hour or up to 8 hours.

❋ Preheat the broiler or grill. Coat a broiler pan or grill rack with no-stick spray. Remove the turkey from the marinade; reserve the marinade.

❋ Broil or grill the turkey 6" from the heat, basting frequently with the reserved marinade, for 15 to 20 minutes, or until no longer pink in the center. Check by inserting the tip of a sharp knife into the thickest part of the breast. Discard any remaining marinade. Cut the turkey into thin slices.

KITCHEN TIP

To freeze, pack the cooled sliced cooked turkey in a freezer-quality plastic container. To use, thaw overnight in the refrigerator. The cold turkey can be used for sandwiches and main-dish salads. To reheat, cover and microwave on high power for 5 minutes, or until hot.

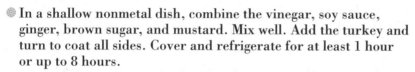

HOMEMADE TURKEY TV DINNERS SAVE BUCKS

Whenever you buy a whole bird, transform the cooked leftovers into individual frozen dinners. Plate the leftover sliced cooked turkey with cooked vegetables, rice, or pasta on sturdy plastic meal trays or plastic picnic plates. Cover with freezer-quality foil, then seal tightly in freezer-quality plastic bags, pressing out as much air as possible. Label the dinners, including the date, and freeze for up to 2 months. Thaw overnight in the refrigerator. To reheat, remove the foil and microwave on high power for 5 to 7 minutes, or until hot. Your money-saving homemade turkey dinners will come in handy on hectic weeknights.

MUSTARDY BAKED TURKEY DRUMSTICKS

*Make these spicy drumsticks on a weekend and freeze for quick weeknight
suppers at less than $1 a serving.*

2 pounds turkey drumsticks (2 drumsticks),
    skinned
¼ cup Dijon mustard
2 tablespoons packed brown sugar
2 cloves garlic, minced
1 teaspoon dry mustard
1 teaspoon hot-pepper sauce
¼ cup dry bread crumbs

✤ Preheat the oven to 400°F. Line a 13" × 9" baking dish with foil.
Place the drumsticks in the dish.

✤ In a small bowl, combine the Dijon mustard, brown sugar,
garlic, dry mustard, and hot-pepper sauce. Brush over the
drumsticks. Sprinkle with the bread crumbs. Bake for 45 minutes, or until no longer pink the center. Check by inserting the
tip of a sharp knife into the thickest part of 1 drumstick. Cut the
meat off of the bones and slice to serve.

MOROCCAN TURKEY WITH CARROTS

A thick paste of herbs and spices coats turkey drumsticks in this Middle Eastern entrée. You can freeze the cooked turkey, but you can also rub the raw meat with the spice mixture and freeze that.

¼ cup minced onions
3 tablespoons lime juice
1 tablespoon minced garlic
1 tablespoon olive oil
1 teaspoon grated fresh ginger
½ teaspoon curry powder
½ teaspoon paprika
¼ teaspoon ground cinnamon
2 pounds turkey drumsticks (2 drumsticks), skinned
2 cups sliced carrots
1 cup apple juice
1 tablespoon packed brown sugar
¼ teaspoon ground cumin

❋ In a 13" × 9" baking dish, combine the onions, lime juice, garlic, oil, ginger, curry powder, paprika, and cinnamon. Mix well. Add the turkey and pat the mixture onto the pieces to coat them evenly. Cover and refrigerate overnight.

❋ Preheat the oven to 400°F.

❋ In a medium bowl, combine the carrots, apple juice, brown sugar, and cumin. Mix well. Spoon around the turkey. Cover and bake for 1 hour, or until the turkey is no longer pink in the center. Check by inserting the tip of a sharp knife into the thickest part of 1 drumstick.

❋ To serve, cut the turkey off of the bone. Slice and serve with the vegetables.

Makes 4 servings

Per serving
Calories 293
Total fat 8 g.
Saturated fat 1.9 g.
Cholesterol 134 mg.
Sodium 140 mg.
Fiber 2 g.

Cost per serving

86¢

KITCHEN TIP

To freeze, pack the cooled cooked turkey and vegetables in a freezer-quality plastic container. To use, thaw overnight in the refrigerator. Place in a 13" × 9" baking dish. Cover and bake at 400°F for 30 to 35 minutes, or until hot.

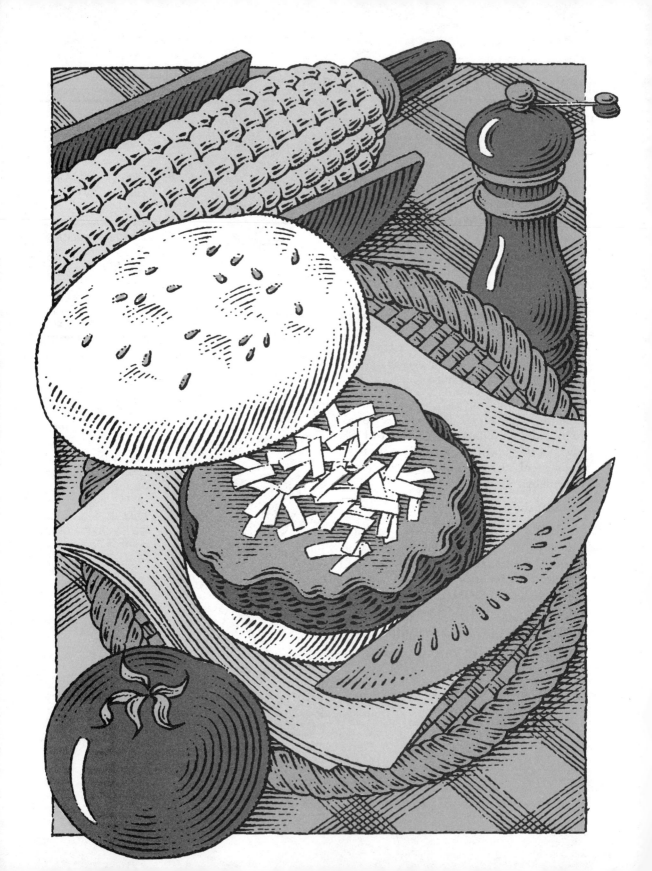

CHAPTER 5

BEEF, PORK, AND LAMB

....

Freezing favorite cuts of beef, pork, and lamb allows you to buy big at supermarket sales. You can stock up when prices fall down. (See "How Long Can I Keep It?" on page 12.)

With a variety of meat on hand, you can cook extra amounts, then store the leftovers in meal-size portions for up to six weeks. Meal preparation goes much faster when you have cooked beef, pork, or lamb to add to hash, stir-fries, sandwiches, pasta dishes, even main-dish salads.

To create convenience dishes, prepare stews and other braised dishes to freeze for up to three months. Cool the cooked dish and then refrigerate overnight before packing for the freezer. Most of the fat in the sauce will rise to the top and harden, making it easy to remove. This technique drastically reduces the fat content of the finished dish.

Variety, nutrition, and convenience are yours with these versatile meat main dishes.

Per serving
Calories 243
Total fat 4.5 g.
Saturated fat 1.2 g.
Cholesterol 31 mg.
Sodium 300 mg.
Fiber 3 g.

Cost per serving

68¢

SLOPPY JOES

Add chopped frozen vegetables to this irresistibly messy sandwich and you cut the cost by 25¢ a serving over the classic ground-meat recipe. You cut down on fat as well. You can make quick work of thawing the ground beef in the microwave just before cooking.

1	cup chopped frozen onions
½	cup chopped frozen green peppers
½	cup chopped frozen carrots
2	tablespoons minced garlic
12	ounces frozen extra-lean ground round beef, thawed
¼	cup reduced-sodium ketchup
1	tablespoon reduced-sodium Worcestershire sauce
1	can (8 ounces) reduced-sodium tomato sauce
½	teaspoon ground cumin
¼	teaspoon ground red pepper
6	reduced-sodium hamburger buns, split and toasted

❋ Coat a 10″ no-stick skillet with no-stick spray and place over medium-high heat until hot. Add the onions, green peppers, carrots, and garlic. Cook, stirring, for 1 minute. Add the beef. Cook, stirring, for 3 minutes, or until the beef begins to brown. Add the ketchup, Worcestershire sauce, tomato sauce, cumin, and red pepper.

❋ Cover and cook, stirring occasionally, for 20 minutes, or until thick. Serve on the buns.

WHERE'S THE FAT?

Fortunately for your waistline and your wallet, the leanest cuts of beef also cost the least. Look for flank steak, top or bottom round, and chuck roast—economical cuts that contain less fat per serving than sirloin. Meat labeled "select," no matter the cut, contains less fat than prime and choice grades. Check if your market carries this grade of meat.

Cooking leaner cuts of beef slowly in a moist environment helps compensate for the smaller amount of fat they contain. To tenderize lean beef, cut it into bite-size pieces and marinate it in a citrus or vinegar marinade made with orange juice, lime juice, lemon juice, cider vinegar, or balsamic vinegar.

Pizza Burgers with Trimmings

Two American all-stars—burgers and pizza—combine in this easy entrée. The burgers can easily be frozen; add the trimmings just before serving.

Burgers

- 1 pound extra-lean ground round beef
- ¼ cup minced onions
- ¼ cup soft bread crumbs
- 2 tablespoons reduced-sodium ketchup
- 2 cloves garlic, minced
- ½ teaspoon dried oregano
- ½ teaspoon dried basil
- ¼ teaspoon ground black pepper

Trimmings

- ¼ cup shredded nonfat mozzarella cheese
- 4 whole-wheat hamburger buns, split and toasted
- 1 teaspoon Dijon mustard
- 1 teaspoon nonfat mayonnaise
- 4 slices tomato

❋ *To make the burgers:* Preheat the broiler. Coat the broiler pan with no-stick spray.

❋ In a medium bowl, combine the beef, onions, bread crumbs, ketchup, garlic, oregano, basil, and pepper. Mix well. Form into 4 patties. Place the patties on the broiler pan. Broil 4" from the heat for 8 minutes. Turn and broil for 5 minutes, or until the beef is no longer pink in the center. Check by inserting the tip of a sharp knife into 1 patty. Do not turn off the broiler.

❋ *To make the trimmings:* Top the burgers with the mozzarella. Broil for 1 minute, or until the cheese melts.

❋ Spread the hamburger buns with the mustard and mayonnaise. Place the burgers on the buns. Top with the tomatoes.

Makes 4 servings

Per serving
Calories 442
Total fat 8.5 g.
Saturated fat 2 g.
Cholesterol 63 mg.
Sodium 613 mg.
Fiber 7.5 g.

Cost per serving

56¢

Kitchen Tip

To freeze, place the cooled cooked burgers on a tray. Put in the freezer for several hours, or until solid. Pack in a freezer-quality plastic bag. To use, thaw overnight in the refrigerator. Reheat in a covered 12" no-stick skillet over medium heat for 5 minutes, or until hot.

Per serving
Calories 176
Total fat 7 g.
Saturated fat 2.2 g.
Cholesterol 77 mg.
Sodium 303 mg.
Fiber 1 g.

Cost per serving

42¢

KITCHEN TIP

To freeze, remove
the cooled cooked
loaf from the pan.
Wrap with a double
layer of freezer-
quality foil. To use,
thaw overnight in
the refrigerator.
Bake the foil-
wrapped loaf
at 350°F for
20 minutes, or
until hot.

MEXICALI MEAT LOAF

*Spicy salsa gives this meat loaf character, and corn adds sweetness. If you
double or triple this recipe to freeze, line the loaf pans with enough over-
hanging freezer-quality foil to wrap the baked loaves for the freezer.*

1	pound extra-lean ground round beef
8	ounces ground turkey breast
¾	cup salsa
¾	cup soft bread crumbs
1	cup chopped onions
½	cup whole kernel corn
1	egg, lightly beaten
2	cloves garlic, minced
1½	teaspoons chili powder
½	teaspoon ground cumin
¼	teaspoon ground black pepper
⅛	teaspoon salt
¼	cup reduced-sodium tomato sauce
1	tablespoon reduced-sodium ketchup
1	teaspoon sugar

✸ Preheat the oven to 350°F.

✸ In a large bowl, combine the beef, turkey, salsa, bread crumbs,
onions, corn, egg, garlic, chili powder, cumin, pepper, and salt.
Mix well. Form into a loaf. Place in a 9" × 5" no-stick loaf pan.

✸ In a small bowl, combine the tomato sauce, ketchup, and sugar.
Spread over the top of the meat loaf. Cover with foil and bake
for 1 hour, or until a meat thermometer inserted into the
thickest part of the meat loaf reads 160°F. Uncover and bake for
5 minutes, or until the top browns slightly. Let stand for 10 minutes before slicing.

BEEF STEW WITH NOODLES

Slow cooking tenderizes the lean chuck roast in this classic dish. Get the most from the cooking time by making a double batch.

1   tablespoon oil
2   pounds lean chuck roast, trimmed of fat
        and cut into 2" cubes
¼   cup apple juice or white wine
2   cups sliced carrots
8   ounces pearl onions, peeled
2   tablespoons minced garlic
2   tablespoons all-purpose flour
4   cups frozen defatted Chicken Stock (page 61),
        thawed
1   teaspoon dried thyme
8   ounces no-yolk noodles

❋ Coat a Dutch oven with no-stick spray and place over medium-high heat. Add the oil and heat until hot. Add the beef. Cook, stirring frequently, for 7 to 10 minutes, or until brown. Transfer to a plate.

❋ Add the apple juice or wine to the Dutch oven. Bring to a boil and scrape the bottom to loosen any browned bits. Add the carrots, onions, and garlic. Cook, stirring, for 5 minutes, or until the onions are soft but not browned. Add the flour. Cook, stirring, for 2 minutes. Add the stock, thyme, and beef. Bring to a boil.

❋ Reduce the heat to medium. Cover and cook for 40 minutes, or until the stew is thick and the beef is no longer pink in the center. Check by inserting the tip of a sharp knife into 1 cube.

❋ Cook the noodles according to the package directions. Drain and serve with the stew.

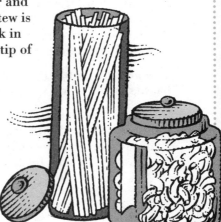

Makes 4 servings

Per serving
Calories 564
Total fat 13 g.
Saturated fat 3.8 g.
Cholesterol 114 mg.
Sodium 152 mg.
Fiber 5.4 g.

Cost per serving

$1.44

KITCHEN TIP

To freeze, pack the cooled cooked stew in a freezer-quality plastic container. Rinse the noodles with cold water after draining and pack in a freezer-quality plastic bag or container. To use, thaw both overnight in the refrigerator. Cover the stew and microwave on high power for 5 minutes, or until hot. Separately microwave the noodles on high power for 3 minutes, or until hot.

Per serving
Calories 251
Total fat 6.5 g.
Saturated fat 3.3 g.
Cholesterol 52 mg.
Sodium 229 mg.
Fiber 2.5 g.

Cost per serving

53¢

KITCHEN TIP

To freeze, wrap the cooled potpie in freezer-quality plastic wrap and then in freezer-quality foil. To use, thaw overnight in the refrigerator. Remove the foil and plastic wrap; discard the plastic wrap. Cover with the foil and bake at 350°F for 20 minutes, or until hot.

VEGETABLE-BEEF POTPIE

Nothing says home cooking like the aroma of bubbling hot potpie. And this biscuit crust couldn't be more foolproof. Just pat with your hands—no rolling required. It freezes well, too, so you can save time by making one pie for dinner and one for the freezer.

Crust

1 cup all-purpose flour
1 teaspoon baking powder
⅛ teaspoon salt
2 tablespoons chilled butter, cut into small pieces
¼ cup skim milk
3 tablespoons nonfat sour cream

Filling

1 small onion, chopped
1 pound extra-lean ground round beef
2 tablespoons cornstarch
2 cups frozen defatted Chicken Stock (page 61), thawed
¾ teaspoon dried thyme
½ teaspoon garlic powder
½ teaspoon Worcestershire sauce
½ cup peas
½ cup sliced carrots

✽ *To make the crust:* In a medium bowl, combine the flour, baking powder, and salt. Mix well. With a pastry blender or fork, cut in the butter until the mixture resembles fine crumbs. Add the milk and sour cream. Mix well. Turn the dough onto a sheet of plastic wrap. Flour your hands and flatten it into a large pancake. Wrap tightly. Refrigerate for 30 minutes.

✽ *To make the filling:* Preheat the oven to 425°F. Coat a 13" × 9" baking dish with no-stick spray.

✽ Coat a 10" no-stick skillet with no-stick spray and place over medium-high heat until hot. Add the onions. Cook, stirring, for 5 minutes, or until the onions are soft but not browned. Add the beef. Cook, stirring, for 5 minutes, or until the beef is browned.

* Place the cornstarch in a medium bowl. Stir in the stock until smooth. Add the thyme, garlic powder, and Worcestershire sauce. Add to the skillet. Bring to a boil. Cook, stirring, for 1 minute. Stir in the peas and carrots. Pour into the prepared baking dish.

* Cut the dough into 4 sections, arrange on top of the filling, with some space between the pieces.

* Bake for 30 minutes, or until the top is golden brown and the filling is bubbling.

BULK UP

Look for bulk buys to save money on meat and stock your freezer. Call local butchers; many give a 10 to 20 percent discount on orders over $50. Some even specialize in selling larger cuts of meat that they buy directly from ranchers.

A good deal in beef is the whole bottom round, which yields good-quality ground round beef, a rump roast, and the more expensive eye of the round. You'll pay about the same price as you would pay for ground round. A butcher will probably charge you about 10¢ a pound to cut and package it.

Another good place to shop for bulk meats is your local warehouse store. Look for specials on frozen pork chops, flank steak, extra-lean ground round beef, or lean ground pork. For instance, you can buy pork loin chops in bulk in the freezer section and save 30¢ a pound over supermarket prices.

Your supermarket may also run specials on these items regularly. Buy as much as your freezer and your budget will allow. Wrap in meal-size portions before freezing.

BEEF AND CABBAGE STIR-FRY

This stir-fry makes an appealing meal for about $1 a serving. For easier slicing of the raw beef, cut while still partially frozen. Serve over noodles or rice.

2   tablespoons reduced-sodium soy sauce
2   cloves garlic, minced
1   tablespoon sugar
2   teaspoons dark sesame oil
1   teaspoon hot-pepper sauce
1   teaspoon cornstarch
1   pound frozen lean chuck roast
¼   cup apple juice
1   cup frozen whole kernel corn
1   cup frozen chopped onions
3   cups shredded green cabbage
1   cup frozen sliced carrots

* In a shallow nonmetal dish, combine the soy sauce, garlic, sugar, oil, hot-pepper sauce, and cornstarch. Mix well. With a serrated knife, slice the beef across the grain into thin strips. Add to the dish and mix well. Cover and refrigerate for 30 minutes, stirring occasionally.

* Coat a 10" no-stick skillet with no-stick spray and place over medium-high heat until hot. Add the beef; reserve the marinade. Cook, stirring, for 5 minutes, or until no longer pink in the center. Check by inserting the tip of a sharp knife into 1 strip. Transfer to a plate; cover to keep warm.

* Add the apple juice to the skillet. Bring to a boil, and scrape the bottom to loosen any browned bits. Add the corn and onions. Cook, stirring, for 3 minutes, or until the onions are soft but not browned. Add the cabbage and carrots. Cook, stirring, for 3 to 4 minutes, or until the vegetables are tender.

* Add the beef and the reserved marinade. Bring to a boil. Cook, stirring, for 1 minute, or until the sauce thickens slightly.

Makes 4 servings

Per serving
Calories 292
Total fat 9.5 g.
Saturated fat 2.9 g.
Cholesterol 79 mg.
Sodium 376 mg.
Fiber 3.7 g.

Cost per serving

$1.02

KITCHEN TIP

For variety, substitute 3 cups frozen broccoli florets for the cabbage.

FREEZE MEAT TENDER

Throw away the meat tenderizer and let the freezer do the work. Add any of the following marinades to 1 pound uncooked lean meat or poultry before freezing. The marinade will not only help break down the tough meat fibers but also season the meat so it's ready to cook straight from the freezer. A marinade should just coat the pieces of food— much like an ideal salad dressing adheres to the lettuce leaves. Butchers recommend marinating for up to 3 months in the freezer. The thinner the cut, the more tenderizing takes place. In a thick cut, the marinade will penetrate only the top 1 to 2 inches.

Cajun. Combine ⅓ cup defatted Chicken Stock (page 61) and ½ teaspoon each of ground red pepper, ground black pepper, ground cinnamon, garlic powder, onion powder, and packed brown sugar.

Citrus. Combine ⅓ cup orange juice and ½ teaspoon each of minced garlic, chopped lemon rind, dried basil, and ground black pepper.

Curry. Combine ⅓ cup defatted Chicken Stock, 1 tablespoon minced garlic, and 1 teaspoon curry powder.

Garlic. Combine ¼ cup defatted chicken stock, 1 tablespoon minced garlic, 1 tablespoon reduced-sodium soy sauce, and 1 teaspoon grated fresh ginger.

Ginger. Combine ¼ cup balsamic vinegar, 2 tablespoons white wine or apple juice, 2 tablespoons grated fresh ginger, and 1 tablespoon minced garlic.

Soy-Apple. Combine ⅓ cup apple juice, 2 tablespoons reduced-sodium soy sauce, and 1 tablespoon defatted Chicken Stock.

Spicy Beef Wraps

Cut the flank steak into strips while it's frozen, then let it thaw right in the lime marinade.

Makes 6 servings

Per serving
Calories 409
Total fat 11.4 g.
Saturated fat 4.2 g.
Cholesterol 59 mg.
Sodium 517 mg.
Fiber 2.4 g.

Cost per serving

$1.70

¼ cup lime juice
2 teaspoons olive oil
4 cloves garlic, minced
1½ teaspoons ground cumin
1½ pounds frozen flank steak, trimmed of fat
2 cups frozen chopped sweet red peppers
1 cup frozen chopped green peppers
2 cups frozen chopped onions
1 cup nonfat plain yogurt
½ medium cucumber, peeled and grated
¼ teaspoon salt
6 frozen pita bread rounds (6" diameter)

❋ In a shallow nonmetal dish, combine the lime juice, oil, garlic, and 1 teaspoon of the cumin. Mix well. With a serrated knife, slice the steak against the grain into thin strips. Add to the dish and mix well. Cover and refrigerate for 1 hour, stirring occasionally.

❋ Coat a 10" no-stick skillet with no-stick spray and place over medium-high heat until hot. Add the red peppers, green peppers, and onions. Cook, stirring, for 5 minutes, or until soft but not browned. Add the steak and marinade. Cook, stirring, for 8 minutes, or until the steak is lightly browned and no longer pink in the center. Check by inserting the tip of a sharp knife into 1 strip.

❋ In a small bowl, combine the yogurt, cucumber, salt, and the remaining ½ teaspoon cumin. Mix well.

❋ Warm the pitas in the microwave on high power for 1 to 2 minutes.

❋ Spoon the steak mixture onto the top of the whole pitas. Top with the yogurt sauce and roll up.

Per serving
Calories 511
Total fat 10.1 g.
Saturated fat 2.9 g.
Cholesterol 79 mg.
Sodium 243 mg.
Fiber 5.1 g.

Cost per serving

93¢

KITCHEN TIP

To freeze, pack the cooled cooked beef and spaghetti in a freezer-quality plastic container. To use, thaw overnight in the refrigerator. Cover and microwave on high power for 5 minutes, or until hot.

CHINESE BEEF AND PASTA SKILLET SUPPER

Chuck roast tenderized in pineapple juice beefs up this easy pasta dish.

1 cup unsweetened pineapple juice
1 tablespoon reduced-sodium soy sauce
1 tablespoon packed brown sugar
1 teaspoon dry mustard
1 pound lean chuck roast, trimmed of fat and cut into 2" cubes
1½ teaspoons dark sesame oil
1 cup diced onions
8 ounces carrots, sliced
1 teaspoon minced garlic
½ teaspoon grated fresh ginger
8 ounces spaghetti
1 teaspoon cornstarch
1 tablespoon water
½ cup minced scallions

✻ In a shallow nonmetal dish, combine the pineapple juice, soy sauce, brown sugar, and mustard. Mix well. Add the beef and mix well. Cover and refrigerate for 1 hour, stirring occasionally.

✻ Coat a 10" no-stick skillet with no-stick spray and place over medium-high heat. Add the oil and heat until hot. Add the onions, carrots, garlic, and ginger. Cook, stirring, for 5 minutes, or until the onions are soft but not browned. Add the beef mixture. Cook, stirring, for 10 minutes, or until the beef is no longer pink in the center. Check by inserting the tip of a sharp knife into 1 cube.

✻ Cook the spaghetti in a large pot of boiling water according to the package directions. Drain well.

✻ Place the cornstarch in a cup. Add the water and stir until smooth. Add to the skillet. Cook, stirring, for 2 to 3 minutes, or until the sauce thickens. Add the spaghetti and scallions; toss well. Cover and cook for 1 minute, or until the spaghetti is hot.

TANGY FLANK STEAK WITH ONIONS

Scoring the surface of the flank steak before marinating it allows more of the marinade to penetrate and tenderize the beef. The sugar in the balsamic vinegar helps caramelize the onions.

½ cup frozen apple juice concentrate
1 tablespoon packed brown sugar
1 teaspoon dry mustard
½ teaspoon paprika
¾ cup balsamic vinegar
1 pound frozen flank steak, trimmed of fat
2 cups sliced frozen onions
1 teaspoon minced garlic
¼ cup frozen defatted Chicken Stock (page 61), thawed

✻ In a shallow nonmetal dish, combine the apple juice concentrate, brown sugar, mustard, paprika, and ½ cup of the vinegar. Mix well. With a sharp knife, score the surface of the steak several times on both sides. Place it the dish, turning to coat. Cover and refrigerate for at least 3 hours or overnight, turning occasionally.

✻ Coat a 10″ no-stick skillet with no-stick spray and place over medium-high heat until hot. Add the onions and garlic. Cook, stirring, for 5 minutes, or until the onions turn golden brown. Add the stock and the remaining ¼ cup vinegar. Cook, stirring, for 8 to 10 minutes, or until the liquid has almost evaporated.

✻ Preheat the broiler. Coat the broiler pan with no-stick spray. Remove the steak from the marinade and place it on the pan. Discard the marinade. Broil the steak 4″ from the heat for 7 minutes. Turn and broil for 7 minutes, or until the steak is no longer pink in the center. Check by inserting the tip of a sharp knife. Serve with the onions.

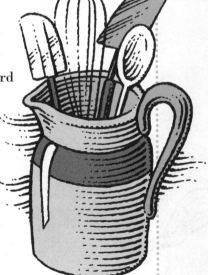

Makes 4 servings

Per serving
Calories 297
Total fat 9.3 g.
Saturated fat 3.9 g.
Cholesterol 59 mg.
Sodium 89 mg.
Fiber 1 g.

Cost per serving

$1.38

KITCHEN TIP

For variety, replace the balsamic vinegar with apple cider vinegar. It harmonizes nicely with the apple juice concentrate.

Per serving
Calories 226
Total fat 8.5 g.
Saturated fat 3 g.
Cholesterol 93 mg.
Sodium 110 mg.
Fiber 0.3 g.

Cost per serving

71¢

KITCHEN TIP

To freeze, slice the
cooled cooked
brisket and pack in
a freezer-quality
plastic container.
To use, thaw
overnight in the
refrigerator. Cover
and microwave
on high power for
5 minutes, or
until hot.

BEEF BRISKET IN CHILI PEPPER SAUCE

This recipe serves 12, which makes it a great make-ahead main dish for a party. Or prepare it for a family dinner and freeze the leftovers for terrific sandwiches. Baking the beef for several hours—while you do something else around the house—makes it fork-tender. The chili pepper sauce translates into zesty pan juices.

½ cup apple juice or red wine
½ cup reduced-sodium tomato sauce
½ cup diced onions
 1 tablespoon packed brown sugar
½ teaspoon ground black pepper
¼ teaspoon salt
 4 pounds beef brisket, trimmed of fat
 1 teaspoon cornstarch
 2 tablespoons water
 1 tablespoon minced jalapeño peppers
     (wear plastic gloves when handling)
¼ teaspoon paprika

❀ Preheat the oven to 350°F.

❀ Coat a 13" × 9" baking dish with no-stick spray. Add the apple juice or wine, tomato sauce, onions, brown sugar, black pepper, and salt. Mix well. Add the beef. Turn to coat all sides.

❀ Cover and bake for 3 hours. Uncover and bake for 20 minutes, or until the beef is tender and no longer pink in the center. Check by inserting the tip of a sharp knife. Transfer to a plate.

❀ Place the cornstarch in a small saucepan. Add the water and stir until smooth. Pour the cooking liquid from the baking dish into the saucepan. Add the jalapeño peppers and paprika. Bring to a boil over medium-high heat. Cook, stirring, for 3 to 5 minutes, or until the gravy thickens. Serve with the brisket.

PORK AND PASTA TETRAZZINI

Makes 4 servings

Per serving
Calories 497
Total fat 8.6 g.
Saturated fat 3.4 g.
Cholesterol 44 mg.
Sodium 354 mg.
Fiber 5.1 g.

Cost per serving

99¢

A creamy sauce and a bounty of vegetables make this casserole comforting and colorful. It's an ideal Sunday supper and a snap to make. For convenience, you may want to cut the pork shoulder into cubes and freeze in 8-ounce portions. This will thaw in just minutes in the microwave.

8 ounces spaghetti
8 ounces frozen pork shoulder, trimmed of fat
  and cut into 1" strips
½ cup frozen chopped sweet red peppers
1 cup frozen pearl onions
1 cup sliced mushrooms
1 cup frozen whole kernel corn
1 cup frozen sliced green beans
2 cloves garlic, minced
¼ cup all-purpose flour
½ teaspoon dried thyme
1½ cups skim milk
¾ cup shredded nonfat sharp Cheddar cheese
¼ cup grated Parmesan cheese
1 teaspoon paprika

❋ Cook the spaghetti in a large pot of boiling water according to the package directions. Drain well and transfer to a large bowl.

❋ Preheat the oven to 350°F. Coat a 13" × 9" baking dish with no-stick spray.

❋ Coat a 10" no-stick skillet with no-stick spray and place over medium-high heat until hot. Add the pork, peppers, and onions. Cook, stirring, for 5 minutes, or until the pork is no longer pink in the center. Check by inserting the tip of a sharp knife into 1 strip.

❋ Add the mushrooms, corn, beans, and garlic. Cook, stirring, for 5 minutes, or until the vegetables soften. Add the flour and thyme. Cook, stirring, for 2 minutes.

❋ Gradually add the milk, stirring constantly. Cook for 2 minutes, or until the sauce thickens. Pour into the bowl with the spaghetti.

❋ Add the Cheddar and Parmesan. Toss well to coat the spaghetti with the sauce. Spoon into the prepared baking dish. Sprinkle with the paprika. Bake for 20 minutes, or until bubbly.

BARBECUED PORK KABOBS WITH COUSCOUS

Barbecue sauce and pork shoulder go together like summer and grilling. This patio supper is easy to love at less than $1.25 a serving.

Juice of 1 lemon
¼ cup reduced-sodium barbecue sauce
1 tablespoon reduced-sodium ketchup
1 teaspoon Dijon mustard
1 teaspoon packed brown sugar
½ teaspoon minced garlic
1 pound pork shoulder, trimmed of fat and
    cut into 1" cubes
½ medium red onion, cut into 1" cubes
1 green pepper, cut into 1" cubes
1 cup frozen defatted Chicken Stock
    (page 61), thawed
1 cup couscous

✱ In a shallow nonmetal dish, combine the lemon juice, barbecue sauce, ketchup, mustard, brown sugar, and garlic. Mix well. Add the pork and stir to coat. Cover and refrigerate for 30 minutes.

✱ Preheat the grill or broiler. Coat the grill rack or broiler pan with no-stick spray. Thread the pork, onions, and peppers onto four 6" metal skewers. Reserve the marinade. Grill or broil 4" from the heat, turning frequently and basting with the reserved marinade, for 5 minutes, or until the pork is no longer pink in the center. Check by inserting the tip of a sharp knife into 1 cube. Discard any remaining marinade.

✱ Bring the stock to a boil in a small saucepan over medium-high heat. Add the couscous and stir well. Cover and remove from the heat. Let stand for 5 minutes. Fluff with a fork. Serve with the pork and vegetables.

Sweet-and-Sour Pork

When you buy pork shoulder on sale, you can make this popular main dish for only $1.99 a serving compared with $3.95 or higher at a Chinese restaurant. Serve it over rice.

¼ cup frozen apple juice concentrate, thawed
1 teaspoon grated fresh ginger
1 teaspoon minced garlic
1 pound pork shoulder, trimmed of fat and
    cut into 1" cubes
1 cup frozen pearl onions, thawed
1 sweet red pepper, diced
1 green pepper, diced
1 tablespoon cornstarch
2 tablespoons reduced-sodium soy sauce
1 tablespoon cider vinegar
1 can (8 ounces) unsweetened pineapple
    chunks (with juice)
1 tablespoon packed brown sugar

❋ In a shallow nonmetal dish, combine the apple juice concentrate, ginger, and garlic. Mix well. Add the pork and stir to coat. Cover and refrigerate for 30 minutes, stirring frequently.

❋ Coat a 10″ no-stick skillet with no-stick spray and place over medium-high heat until hot. Add the onions, red peppers, and green peppers. Cook, stirring, for 3 minutes, or until crisp-tender. Add the pork and marinade. Cook, stirring, for 5 minutes, or until the pork is no longer pink in the center. Check by inserting the tip of a sharp knife into 1 cube.

❋ Place the cornstarch in a medium bowl. Add the soy sauce and vinegar. Stir until smooth. Stir in the pineapple (with juice) and brown sugar. Add to the skillet. Bring to a boil. Cook, stirring frequently, for 3 minutes, or until the sauce thickens.

Makes 4 servings

Per serving
Calories 290
Total fat 10.6 g.
Saturated fat 3.7 g.
Cholesterol 69 mg.
Sodium 369 mg.
Fiber 1.1 g.

Cost per serving

$1.99

Kitchen Tip

To freeze, pack the cooled cooked pork in a freezer-quality plastic container. To use, thaw overnight in the refrigerator. Cover and microwave on high power for 5 minutes, or until hot.

PACK PORK PROPERLY

If you get a terrific buy on pork tenderloin, freeze it properly to keep its flavor as fresh as the day you bought it.

Package uncooked pork tenderloin in freezer-quality plastic bags, plastic wrap, or foil. Wrap tightly to prevent moisture loss. Label and then freeze for up to six months. Thaw overnight in the refrigerator before using in a recipe.

If you prefer to cook the pork tenderloin in a dish before freezing, remember that dishes with a sauce or gravy keep the meat moist and flavorful. Pack the completely chilled, cooked dish in a freezer-quality plastic container that holds the food with about ½ inch of space between the food and the lid to allow for expansion. Don't leave too much room, however, because larger pockets of air make a nice home for ice crystals to form. That can give your food unwelcome freezer burn, which sours the taste and harms the texture of the dish.

Packaged correctly, cooked pork dishes can be frozen for up to three months without loss of flavor. When you're ready to serve, thaw the dish overnight in the refrigerator. Cover and microwave on high power for 5 minutes, or until hot.

CITRUS-MARINATED PORK CHOPS

Makes 4 servings

Per serving
Calories 188
Total fat 5.7 g.
Saturated fat 1.9 g.
Cholesterol 43 mg.
Sodium 393 mg.
Fiber 0.7 g.

Cost per serving

$1.39

Here's a case where you can prepare the marinade, add the uncooked chops, and freeze until needed. The flavor of the marinade actually intensifies during the freezing time.

     Juice of 2 limes
     Juice of 2 oranges
2 tablespoons balsamic vinegar
2 tablespoons honey
1 tablespoon Dijon mustard
1 tablespoon paprika
1 teaspoon minced garlic
1 teaspoon ground black pepper
½ teaspoon salt
4 pork loin chops, trimmed of fat

✷ In a shallow nonmetal dish, combine the lime juice, orange juice, vinegar, honey, mustard, paprika, garlic, pepper, and salt. Mix well. Add the pork and turn to coat. Cover and refrigerate for 30 minutes, turning occasionally.

✷ Coat a 10″ no-stick skillet with no-stick spray and place over medium-high heat until hot. Add the pork; reserve the marinade. Cook for 5 minutes, or until brown. Turn and cook for 5 minutes, or until brown.

✷ Add the marinade. Reduce the heat to medium. Cover and cook for 7 minutes, or until the pork is no longer pink in the center. Check by inserting the tip of a sharp knife into 1 chop.

✷ Transfer the pork to a plate. Cover to keep warm. Increase the heat to high. Cook, stirring frequently, for 3 to 5 minutes, or until the marinade is thick.
Spoon over the pork.

KITCHEN TIP

To freeze, pack the uncooked pork chops in a freezer-quality plastic container. Add the marinade. To use, thaw overnight in the refrigerator. Cook according to the recipe directions.

ROSEMARY PORK

Rosemary's assertive piney quality is a plus in this dish because it holds its flavor even when frozen.

1 cup chopped onions
2 cloves garlic, minced
1½ teaspoons crushed dried rosemary
1 pound pork shoulder, trimmed of fat and cut into 1" cubes
¼ cup frozen defatted Chicken Stock (page 61), thawed
2 cups chopped cabbage
1 cup chopped carrots
1 cup chopped mushrooms
1 cup apple juice
¼ teaspoon salt
¼ teaspoon ground black pepper
1 cup long-grain white rice
¼ cup chopped scallions

❋ In a medium bowl, combine the onions, garlic, and rosemary. Mix well. Add the pork and stir to coat.

❋ Coat a 10" no-stick skillet with no-stick spray and place over medium-high heat until hot. Add the pork mixture. Cook, stirring, for 5 minutes, or until the pork is browned. Transfer the pork mixture to a plate.

❋ Add the stock to the skillet. Bring to a boil, and scrape the bottom to loosen any browned bits. Add the cabbage, carrots, and mushrooms. Cook, stirring, for 3 minutes. Add the apple juice, salt, pepper, and pork mixture. Cover and cook for 30 minutes, or until the pork is no longer pink in the center. Check by inserting the tip of a sharp knife into 1 cube.

❋ Cook the rice according to the package directions. Toss with the scallions. Serve the pork with the rice.

Jamaican Pork Tenderloin

Pounding slices of pork tenderloin into thin scaloppine makes portions look really abundant. The delightfully fruity sauce keeps the pork moist in the freezer. Serve the pork over rice.

1   **pound pork tenderloin, trimmed of fat**
1   **cup orange juice**
1   **cup chopped onions**
1   **green pepper, diced**
1   **sweet red pepper, diced**
1   **teaspoon cornstarch**
2   **tablespoons frozen apple juice concentrate, thawed**
2   **teaspoons minced garlic**
½   **teaspoon ground red pepper**
¼   **teaspoon ground cumin**
¼   **teaspoon salt**
¼   **cup minced scallions**

✷ Cut the pork crosswise into 16 slices. Place the slices, several at a time, between 2 sheets of wax paper. Using a meat mallet, pound to ¼" thickness.

✷ Coat a 10" no-stick skillet with no-stick spray and place over medium-high heat until hot. Add enough pork slices to cover the pan. Cook for 3 minutes. Turn and cook for 3 minutes, or until browned. Transfer the pork to a plate; cover to keep warm. Repeat until all the pork is cooked.

✷ Add the orange juice to the skillet. Bring to a boil, scraping to loosen any browned bits from the bottom. Add the onions, green peppers, and sweet red peppers. Cook, stirring, for 5 minutes, or until the vegetables soften.

✷ Place the cornstarch in a small bowl. Add the apple juice concentrate and stir until smooth. Add the garlic, ground red pepper, cumin, and salt. Add to the skillet. Cook, stirring, for 3 minutes, or until the sauce thickens. Serve over the pork. Sprinkle with the scallions.

Makes 4 servings

Per serving
Calories 210
Total fat 4.5 g.
Saturated fat 1.5 g.
Cholesterol 67 mg.
Sodium 187 mg.
Fiber 1.3 g.

Cost per serving

$1.85

Kitchen Tip

To freeze, pack the cooled cooked pork and sauce in a freezer-quality plastic container. To use, thaw overnight in the refrigerator. Cover and microwave on high power for 5 minutes, or until hot.

Per serving
Calories 273
Total fat 4.8 g.
Saturated fat 1.5 g.
Cholesterol 60 mg.
Sodium 701 mg.
Fiber 1.9 g.

Cost per serving

$1.41

KITCHEN TIP

Get full value from
any bunch of herbs
you purchase.
After using the
amount you need
for a recipe, chop
and freeze the
remaining leaves in
a small freezer-
quality plastic bag.
Rosemary and
other herbs will
remain green and
flavorful. Use the
herbs while still
frozen.

STUFFED PORK TENDERLOIN

*With pork tenderloin, onions, and chicken stock in the freezer, you can
pull together this down-home dish by just adding packaged dry cornbread
stuffing, raisins, and some spices.*

2	cups frozen defatted Chicken Stock (page 61), thawed
¼	cup frozen minced onions
2	tablespoons raisins
1½	cups dry cornbread stuffing
1	teaspoon poultry seasoning
2	frozen pork tenderloins (12 ounces each), thawed and trimmed of fat
1	teaspoon chopped fresh rosemary
½	teaspoon ground allspice
¼	teaspoon ground black pepper
2	tablespoons balsamic vinegar
1	teaspoon cornstarch
1	tablespoon water

❋ Bring 1 cup of the stock to a boil in a 10″ no-stick skillet over
medium-high heat. Add the onions and raisins. Cook, stirring
frequently, for 5 minutes, or until the onions are soft. Remove
from the heat. Add the cornbread stuffing and poultry seasoning. Mix well.

❋ Preheat the oven to 425°F.

❋ With a sharp knife, make a lengthwise cut in each pork tenderloin almost, but not completely, through. Open one tenderloin
like a book and place between 2 sheets of wax paper. Using a
meat mallet, pound to ½″ thickness. Repeat with the second tenderloin.

❋ Spoon the stuffing onto each pork tenderloin and roll carefully
to enclose the filling. Tie in several places with kitchen twine or
secure with small skewers.

❋ In a small bowl, combine the rosemary, allspice, and pepper.
Rub over the outside of the tenderloins.

❋ Coat a 10″ no-stick skillet with no-stick cooking spray and place
over medium-high heat until hot. Add the pork. Cook for 2 to 3
minutes, or until brown. Carefully turn and cook for 2 to 3 minutes, or until brown.

- Add the remaining 1 cup stock. Reduce the heat to low. Cover and cook for 25 minutes, or until the pork is tender and cooked through. Check by inserting the tip of a sharp knife into the center of 1 tenderloin.

- Place the pork on a serving platter. Remove the twine or skewers. Slice the pork into ½" slices. Cover to keep warm.

- Add the vinegar to the skillet. Bring to a boil, scraping to loosen any browned bits from the bottom.

- Place the cornstarch in a cup. Add the water and stir until smooth. Add to the skillet. Cook, stirring, until the gravy has thickened. Pour over the pork.

BRAISE FOR PRAISE

Economical cuts of meat are often lean, which makes them on the tough side. But slow cooking will tenderize and improve their flavor. Use the braising method, which is slow cooking in a small amount of liquid in a covered pan. This works well for chuck roast, brisket, and pork shoulder. Braised meat dishes that are frozen are even more succulent after they come out of the deep freeze. As the meat gravy freezes and expands, it helps break down the tougher fibers in lean meat, resulting in morsels that are fork-tender.

Per serving
Calories 263
Total fat 7.9 g.
Saturated fat 2.4 g.
Cholesterol 43 mg.
Sodium 545 mg.
Fiber 1.7 g.

Cost per serving

$1.06

KITCHEN TIP

To freeze, pack the cooled cooked chops in a freezer-quality plastic container. To use, thaw overnight in the refrigerator. Cover and microwave on high power for 5 minutes, or until hot.

SPICY BAKED PORK CHOPS

Orange juice tenderizes these pork chops, and the spicy herb coating gives them zing. The cooked pork chops freeze beautifully with their herb coating for up to 2 months. Serve with rice, tossed green salad, and a fruit dessert.

1   cup orange juice
3   cloves garlic, minced
2   tablespoons reduced-sodium soy sauce
1   teaspoon olive oil
4   pork loin chops, trimmed of fat
1   cup dry bread crumbs
2   teaspoons dried thyme
2   teaspoons paprika
½   teaspoon ground red pepper
¼   teaspoon ground black pepper

✺ In a shallow nonmetal dish, combine the orange juice, garlic, soy sauce, and oil. Mix well. Add the pork and turn to coat. Cover and refrigerate for at least 1 hour or overnight.

✺ In a medium bowl, combine the bread crumbs, thyme, paprika, red pepper, and black pepper.

✺ Preheat the oven to 450°F. Coat a large baking sheet with no-stick spray.

✺ Drain the marinade from the pork; discard the marinade. Add the pork to the bread-crumb mixture and toss well to coat. Place the pork on the baking sheet. Coat the pork with no-stick spray. Bake for 15 to 20 minutes, or until no longer pink in the center. Check by inserting the tip of a sharp knife into 1 chop.

ROAST PORK WITH SWEET POTATOES AND APPLES

This pork roast makes 10 servings, so there are plenty of leftovers to package and freeze. Sweet potatoes, apples, and honey create a hearty flavor, and slow cooking tenderizes the lean meat.

1   boneless pork roast (4 pounds), trimmed of fat
⅓   cup honey
¼   cup orange juice
¼   cup frozen apple juice concentrate, thawed
1   tablespoon ground black pepper
1   tablespoon packed brown sugar
3   large sweet potatoes, halved
3   large apples, cored and quartered

❉ Preheat the oven to 375°F.

❉ Coat a large Dutch oven with no-stick spray and place over medium-high heat until hot. Add the pork. Cook for 2 minutes, or until brown. Turn and cook for 2 minutes, or until brown. Remove from the heat.

❉ In a medium bowl, combine the honey, orange juice, apple juice concentrate, pepper, and brown sugar. Spoon over the pork. Place the sweet potatoes around the pork. Cover and bake for 2 hours, or until the pork has an internal temperature of 150°F.

❉ Place the apples around the pork. Bake, uncovered and basting frequently, for 20 minutes, or until the apples are just tender. Let the pork stand for 10 minutes before slicing. Serve with the sweet potatoes and apples.

Makes 10 servings

Per serving
Calories 351
Total fat 6.9 g.
Saturated fat 2.3 g.
Cholesterol 108 mg.
Sodium 83 mg.
Fiber 2.5 g.

Cost per serving

$1.05

KITCHEN TIP

To freeze, pack the cooled cooked pork, sweet potatoes, and apples in a freezer-quality plastic container. To use, thaw overnight in the refrigerator. Cover and microwave on high power for 5 minutes, or until hot.

KITCHEN TIP

To freeze, pack the
cooled cooked
meatballs and
gravy in a freezer-
quality plastic
container. To use,
thaw overnight in
the refrigerator.
Place in a no-stick
skillet. Cover
and cook over
medium heat for
5 minutes, or
until hot.

SPICY LAMB MEATBALLS WITH PAN GRAVY

You can make these meatballs up to 3 months ahead because they freeze so nicely in the sauce.

12 ounces lean ground lamb
¾ cup dry bread crumbs
¼ cup shredded nonfat mozzarella cheese
1 tablespoon nonfat plain yogurt
1 teaspoon minced garlic
1 teaspoon chopped fresh mint
¼ teaspoon ground red pepper
2 cups frozen defatted Chicken Stock
   (page 61), thawed
1 teaspoon cornstarch
¼ teaspoon salt
¼ teaspoon ground black pepper

✤ In a medium bowl, combine the lamb, bread crumbs, mozzarella, yogurt, garlic, mint, and red pepper. Mix well. Form into twelve 2″ balls.

✤ Coat a 10″ no-stick skillet with no-stick spray and place over medium-high heat until hot. Add the meatballs. Cook for 5 minutes. Turn and cook for 5 minutes, or until browned.

✤ Add 1 cup of the stock. Bring to a boil. Cover and cook for 15 minutes, or until the meatballs are no longer pink in the center. Check by inserting the tip of a sharp knife into 1 meatball.

✤ Transfer the meatballs to a plate. Cover to keep warm.

✤ Place the cornstarch in a small bowl. Add the remaining 1 cup stock and stir until smooth. Stir in the salt and black pepper. Add to the skillet. Bring to a boil, scraping to loosen any browned bits from the bottom. Cook, stirring frequently, for 3 minutes, or until the gravy thickens. Serve over the meatballs.

SHEPHERD'S PIE

Some versions of shepherd's pie are topped with fat-saturated biscuits, but this lean and economical casserole calls for a crown of creamy mashed potatoes to slash the fat and cost.

2 pounds baking potatoes, peeled and cubed
⅓ cup low-fat buttermilk
¼ cup grated Parmesan cheese
1 pound frozen ground lamb, thawed
1 cup frozen pearl onions
1 cup frozen sliced carrots
1 cup frozen chopped cauliflower
2 cloves garlic, minced
1 can (8 ounces) reduced-sodium tomatoes, chopped (with juice)
1 cup frozen peas
2 teaspoons reduced-sodium Worcestershire sauce
1 teaspoon dried Italian herb seasoning
1 teaspoon cornstarch
¼ cup apple juice

* Preheat the oven to 350°F. Coat a 13″ × 9″ baking dish with no-stick spray.

* Place the potatoes in a large saucepan. Add cold water to cover by about 1″. Bring to a boil over high heat. Reduce the heat to medium and cook for 15 minutes, or until very tender. Drain and place in a medium bowl. Mash well with a hand masher or an electric mixer. Add the buttermilk and Parmesan. Mix well.

* Coat a 10″ no-stick skillet with no-stick spray and place over medium-high heat until hot. Add the lamb, onions, carrots, cauliflower, and garlic. Cook, stirring, for 5 minutes, or until the lamb is browned. Add the tomatoes (with juice), peas, Worcestershire sauce, and Italian herb seasoning. Bring to a boil.

* Place the cornstarch in a cup. Add the apple juice and stir until smooth. Add to the skillet. Cook, stirring, for 3 minutes, or until the sauce thickens.

* Spoon into the prepared baking dish. Top with the mashed potatoes. Bake for 40 minutes, or until the topping is golden brown. Let stand for 10 minutes before serving.

Makes 6 servings

Per serving
Calories 283
Total fat 6 g.
Saturated fat 2 g.
Cholesterol 47 mg.
Sodium 185 mg.
Fiber 5.1 g.

Cost per serving

$1.03

Make Ends Meat

You can save big bucks on your grocery bill if you learn to stretch your meat dollar. Here are some tips.

🔸 Eat one, rather than two, meat meals a day. You can save $2,000 a year for a family of four on this strategy alone. Replace meat-centered dinners or brown-bag lunches with ones based on grains and vegetables.

🔸 Always shop specials, then plan your meat meal around what's on sale.

🔸 Buy less expensive cuts. Dark-meat chicken in family packs costs only 22¢ per pound compared with chicken breasts on the bone at 69¢ per pound.

🔸 Use less meat in a recipe. Make chili, spaghetti sauce, soups, and stews with ½ pound of meat instead of a full pound. Make up the difference with beans. Beans are so satisfying—and incredibly low in fat—that no one will know the difference.

🔸 Substitute ground turkey breast for all or part of the red meat in ground beef recipes. It costs less, is generally much lower in fat, and tastes terrific.

🔸 Buy meat in season. Strange as it sounds, there is a seasonality to some meat prices. Steaks are more expensive during the summer (grilling time), and pork is often cheaper in the winter. Use a price book (see page 5), to chart the seasonal changes of meat sales in your area.

LAMB CURRY

Cooks in India designed this dish to use up leftover lamb by combining it with a rainbow of spices and vegetables.

1   medium eggplant, thinly sliced
1   onion, diced
⅓   cup apple juice
1   teaspoon minced garlic
1   pound lamb, trimmed of fat and
       cut into 1" cubes
1   can (14 ounces) reduced-sodium whole
       tomatoes (with juice), chopped
¼   cup frozen defatted Chicken Stock (page 61),
       thawed
3   tablespoons curry powder
1   teaspoon ground black pepper
½   teaspoon ground coriander
½   cup nonfat plain yogurt
2   tablespoons minced fresh cilantro

✻ Coat a 10" no-stick skillet with no-stick spray and place over medium-high heat until hot. Add the eggplant, onions, apple juice, and garlic. Cook, stirring, for 5 minutes. Add the lamb, tomatoes (with juice), stock, curry powder, pepper, and coriander. Bring to a boil.

✻ Reduce the heat to medium. Cook, stirring occasionally, for 20 minutes, or until the curry is thick and the lamb is no longer pink in the center. Check by inserting the tip of a sharp knife into 1 cube.

✻ Serve topped with the yogurt and sprinkled with the cilantro.

Makes 4 servings

Per serving
Calories 248
Total fat 7.7 g.
Saturated fat 2.5 g.
Cholesterol 65 mg.
Sodium 94 mg.
Fiber 6.5 g.

Cost per serving

$1.37

KITCHEN TIP

To freeze, pack the cooled cooked curry in a freezer-quality plastic container. To use, thaw overnight in the refrigerator. Cover and microwave on high power for 5 minutes, or until hot.

CHAPTER 6

FISH AND SHELLFISH

The frozen fish case of the local supermarket is an exiting place to be these days. Farm-raised fish and ocean fish that are flash-frozen and vacuum-packed right on the boat are bringing increasing variety to a wider audience. Frozen fish can often taste far superior to unfrozen fish that has been out of the water for more than a week before it reaches your table.

Fish and shellfish add lean protein and heart-healthy omega-3 fatty acids to your meals. And despite the pricey reputation, seafood can get into the swim of the low-cost kitchen. Scout supermarket sales for the discounted catch of the day and buy your favorite fish in bulk freezer packages at your local warehouse store. You can reduce the price to as low as $1.29 per pound for some varieties.

With smart shopping, careful freezer storage, and the delightful recipes in this chapter, seafood can make a real splash in your kitchen.

Per serving
Calories 379
Total fat 4.9 g.
Saturated fat 0.9 g.
Cholesterol 53 mg.
Sodium 160 mg.
Fiber 2.4 g.

Cost per serving

$1.90

KITCHEN TIP

To freeze, pack the
cooled cooked
flounder, rice, and
vegetables in a
freezer-quality
plastic container.
To use, thaw
overnight in the
refrigerator. Cover
and microwave
on high power for
3 to 5 minutes,
or until hot.

CARIBBEAN FLOUNDER WITH RICE

Flounder is transformed from mild to wild with a jalapeño pepper–spiked accompaniment of tomatoes, scallions, sweet red peppers, and rice.

1	tablespoon olive oil
1	cup chopped scallions
1	cup chopped tomatoes
1	cup long-grain white rice
1	large sweet red pepper, chopped
1	small jalapeño pepper, seeded and chopped (wear plastic gloves when handling)
½	cup frozen defatted Chicken Stock (page 61), thawed
¼	cup cider vinegar
1	teaspoon sugar
½	teaspoon dried thyme
4	flounder fillets (4 ounces each)
⅛	teaspoon salt
⅛	teaspoon ground black pepper

❀ Coat a Dutch oven with no-stick spray and place over medium-high heat until hot. Add the oil, scallions, tomatoes, rice, red peppers, and jalapeño peppers. Cook, stirring, for 5 minutes, or until the peppers are soft. Add the stock, vinegar, sugar, and thyme. Cook, stirring, for 1 minute.

❀ Cover and cook for 15 minutes. Lay the flounder on top of the rice and vegetables. Cover and cook for 10 minutes, or until the rice is tender and the fish is opaque in the center. Check by inserting the tip of a sharp knife into the center of 1 fillet. Sprinkle with the salt and pepper.

SOMETHING'S FISHY

The old adage, "After three days, guests and fish stink," is pretty true for finny creatures. Fish spoils faster than any other protein. The moment a fish is caught, internal enzymes go to work breaking down the flesh.

Savvy cooks rinse and pat dry any fresh fish they buy at the supermarket. After the surface is cleaned, fish will smell fresher and taste better, too. If you don't use your bargain buys on the day of purchase, package and freeze them right away.

Flash-freeze fillets in convenient individual portions. Place the fillets in a single layer on a tray. Place in the freezer for 1 to 2 hours, or until almost frozen. Layer the fillets between pieces of wax paper, then place in a freezer-quality plastic bag, pressing out as much air as possible. Return to the freezer. Make sure your freezer is set for 0°F or colder. Fish with higher fat content, such as salmon, can be frozen for up to three months. Leaner fish, such as halibut, can be frozen successfully for up to six months.

CAJUN CATFISH

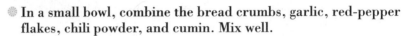

We've coated economical farm-raised catfish with a spicy crust before broiling it. Serve with a tossed green salad, bread, and green beans.

Per serving
Calories 222
Total fat 10.7 g.
Saturated fat 2.4 g.
Cholesterol 82 mg.
Sodium 160 mg.
Fiber 0.4 g.

Cost per serving

$1.52

¼ cup dry bread crumbs
1 teaspoon minced garlic
½ teaspoon crushed red-pepper flakes
½ teaspoon chili powder
½ teaspoon ground cumin
4 catfish fillets (6 ounces each)

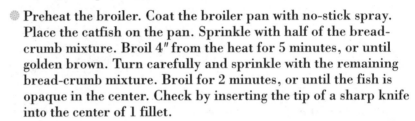

In a small bowl, combine the bread crumbs, garlic, red-pepper flakes, chili powder, and cumin. Mix well.

Preheat the broiler. Coat the broiler pan with no-stick spray. Place the catfish on the pan. Sprinkle with half of the bread-crumb mixture. Broil 4" from the heat for 5 minutes, or until golden brown. Turn carefully and sprinkle with the remaining bread-crumb mixture. Broil for 2 minutes, or until the fish is opaque in the center. Check by inserting the tip of a sharp knife into the center of 1 fillet.

KITCHEN TIP

To freeze, place the cooled cooked catfish on a tray. Put in the freezer for several hours, or until solid. Transfer to a freezer-quality plastic bag. To use, thaw overnight in the refrigerator. Place in a 12" no-stick skillet, cover, and cook over low heat for 8 to 10 minutes, or until hot.

COD IN SWEET TOMATO SAUCE

Slow-cooking the tomatoes creates a supersweet sauce for the cod in this simple one-pot main dish.

2   teaspoons olive oil
2   cups frozen chopped onions
1   cup frozen chopped sweet red peppers
1   cup chopped celery
2   cups chopped tomatoes
2   teaspoons minced garlic
1   teaspoon sugar
¼   cup frozen defatted Fish Stock (page 60), thawed
¼   teaspoon salt
¼   teaspoon ground black pepper
4   frozen cod fillets (6 ounces each)

❋ Coat a Dutch oven with no-stick spray and place over medium-high heat until hot. Add the oil, onions, red peppers, and celery. Cook, stirring, for 5 minutes, or until the onions are soft.

❋ Add the tomatoes, garlic, and sugar. Reduce the heat to medium, cover, and cook, stirring frequently, for 10 minutes, or until the tomatoes soften. Add the stock, salt, and black pepper. Bring to a boil.

❋ Place the cod on top of the tomato sauce. Cover and cook for 10 to 12 minutes, or until the fish is opaque in the center. Check by inserting the tip of a sharp knife into the center of 1 fillet.

Makes 4 servings

Per serving
Calories 253
Total fat 4 g.
Saturated fat 0.6 g.
Cholesterol 80 mg.
Sodium 271 mg.
Fiber 3.5 g.

Cost per serving

$1.60

Per serving
Calories 158
Total fat 3.4 g.
Saturated fat 0.6 g.
Cholesterol 60 mg.
Sodium 250 mg.
Fiber 1.2 g.

Cost per serving

$1.94

KITCHEN TIP

To freeze, pack the cooled cooked cod in a freezer-quality plastic container. To use, thaw overnight in the refrigerator. Cover and microwave on high power for 3 to 5 minutes, or until hot.

COD WITH SALSA VERDE

Baked cod stays moist and flavorful when topped with a sprightly green salsa.

½ cup chopped scallions
½ cup chopped fresh parsley
¼ cup minced onions
¼ cup lemon juice
2 tablespoons capers, rinsed and chopped
4 cloves garlic, minced
2 teaspoons olive oil
2 teaspoons minced jalapeño peppers
     (wear plastic gloves when handling)
4 cod fillets (8 ounces each)

❈ Preheat the oven to 400°F. Place a 24″ × 24″ piece of foil on a baking sheet.

❈ In a small bowl, combine the scallions, parsley, onions, lemon juice, capers, garlic, oil, and peppers. Mix well.

❈ Arrange the cod on the foil in a single layer. Top with the scallion mixture. Fold the sides of the foil over the fish to create a sealed packet.

❈ Bake for 10 minutes, or until the fish is opaque in the center. Check by carefully opening 1 packet and inserting the tip of a sharp knife into 1 fillet.

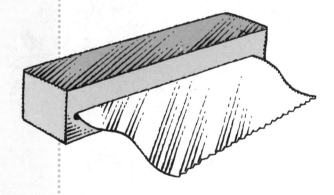

Halibut in Sicilian Sauce

This simple stove-top supper is easy on the eye and the budget. The savory southern Italian sauce steams the fish as it permeates it with flavor.

1 teaspoon olive oil
3 cups chopped onions
1 sweet red pepper, diced
3 tablespoons raisins
2 tablespoons white-wine vinegar
1 teaspoon sugar
2 teaspoons minced garlic
1 teaspoon dried oregano
1 drop hot-pepper sauce
½ teaspoon salt
4 halibut steaks (6 ounces each)
2 tablespoons chopped fresh parsley

✵ Coat a 10″ no-stick skillet with no-stick spray and place over medium heat until hot. Add the oil, onions, and red peppers. Cook, stirring frequently, for 15 minutes, or until the onions are golden brown and very soft. Add the raisins, vinegar, sugar, garlic, oregano, hot-pepper sauce, and salt. Stir well.

✵ Place the halibut on top of the vegetable mixture. Cover and cook for 8 minutes, or until the fish is opaque in the center. Check by inserting the tip of a sharp knife into 1 fillet. Serve sprinkled with the parsley.

Makes 4 servings

Per serving
Calories 258
Total fat 4.9 g.
Saturated fat 0.7 g.
Cholesterol 49 mg.
Sodium 374 mg.
Fiber 3.4 g.

Cost per serving

$1.99

Kitchen Tip

To freeze, pack the cooled cooked halibut in a freezer-quality plastic container. To use, thaw overnight in the refrigerator. Cover and microwave on high power for 3 to 5 minutes, or until hot.

MEDITERRANEAN TUNA AND BEAN SALAD

A traditional combination in the Italian region of Tuscany, this salad makes wonderful use of frozen vegetables and canned tuna for less than 70¢ a serving. You can pull this quick light dish together from staples that you have on hand. If you freeze your cooked beans in 1-cup packages, they'll thaw quickly in the microwave.

Per serving
Calories 214
Total fat 3 g.
Saturated fat 0.6 g.
Cholesterol 18 mg.
Sodium 249 mg.
Fiber 10.5 g.

Cost per serving

69¢

1   cup frozen cooked navy beans
       (page 177), thawed
3   tablespoons balsamic vinegar
1   tablespoon olive oil
1   teaspoon Dijon mustard
2   cloves garlic, minced
2   cups frozen sliced green beans
2   cups frozen broccoli florets
1   cup frozen cauliflower florets
1   cup frozen chopped onions
8   leaves romaine lettuce
1   can (7 ounces) water-packed tuna,
       drained and flaked

In a medium bowl, combine the navy beans, vinegar, oil, mustard, and garlic. Toss well. Let stand at room temperature for 30 minutes, stirring occasionally.

Bring a large pot of water to a boil over medium-high heat. Add the green beans, broccoli, and cauliflower. Cook for 30 seconds, or until the beans turn bright green. Drain in a colander. Add the onions. Rinse under cold water. Pat dry with paper towels. Add to the bowl with the beans. Toss well. Let stand at room temperature for 5 minutes.

Line 4 salad plates with the lettuce. Arrange the vegetable mixture on the lettuce. Scatter the tuna over the lettuce.

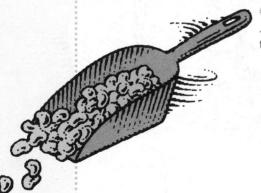

OVEN-FRIED ORANGE ROUGHY

Adding only shelf staples, you can create this satisfying oven-baked entrée in minutes. For an accompaniment, microwave sweet potatoes, then drizzle lightly with honey and sprinkle with cinnamon. In addition, cook frozen green beans in a covered pan with 2 tablespoons Chicken Stock (page 61) and a sprinkling of dried oregano until hot; drizzle with some lemon juice before serving.

¼ cup dry bread crumbs
1 tablespoon grated Parmesan cheese
½ teaspoon baking powder
¼ teaspoon dried thyme
¼ teaspoon dried marjoram
⅛ teaspoon ground red pepper
⅛ teaspoon salt
4 frozen orange roughy fillets (6 ounces each),
   thawed
¼ cup all-purpose flour
¼ cup low-fat buttermilk

❋ Preheat the oven to 450°F. Coat a large baking sheet with no-stick spray.

❋ In a medium bowl, combine the bread crumbs, Parmesan, baking powder, thyme, marjoram, red pepper, and salt. Mix well.

❋ Pat the orange roughy dry with paper towels. Place the flour on a plate. Place the buttermilk in a shallow bowl. Dip the fish into the flour, then into the buttermilk, then into the bread-crumb mixture. Place on the prepared baking sheet. Coat the fish with no-stick spray.

❋ Bake for 10 to 12 minutes, or until the fish is crisp, golden brown, and opaque in the center. Check by inserting the tip of a sharp knife into 1 fillet.

Makes 4 servings

Per serving
Calories 181
Total fat 2.2 g.
Saturated fat 0.5 g.
Cholesterol 35 mg.
Sodium 330 mg.
Fiber 0.6 g.

Cost per serving

$1.99

PAN-FRIED TROUT

Makes 4 servings

Per serving
Calories 264
Total fat 6.4 g.
Saturated fat 1.7 g.
Cholesterol 128 mg.
Sodium 192 mg.
Fiber 3 g.

Cost per serving

$1.61

Buy whole trout on sale and freeze for up to 3 months. Just remember to transfer the trout to the refrigerator the night before cooking so that it thaws properly. Wedges of fresh lemon, if you have them on hand, can be squeezed over the trout just before serving.

1½ cups rolled oats
¼ cup skim milk
2 eggs, lightly beaten
½ cup all-purpose flour
½ teaspoon ground black pepper
¼ teaspoon salt
2 frozen whole trout (12 ounces each), thawed

⊛ In a blender or food processor, grind the oats to a fine powder. Place in a shallow bowl. In a second bowl, mix the milk and eggs. In a third bowl, mix the flour, pepper, and salt.

⊛ Rinse the trout with cold water and pat dry with paper towels. Dip into the flour mixture, then into the egg mixture, then into the oats. Coat with no-stick spray.

⊛ Coat a 10" no-stick skillet with no-stick spray and place over medium-high heat until hot. Add the fish. Cook for 5 minutes, or until golden brown. Turn and cook for 5 to 8 minutes, or until the fish is opaque in the center. Check by inserting the tip of a sharp knife into the center of 1 trout.

⊛ To serve, cut the trout lengthwise down to the backbone. Remove the top 2 fillets to dinner plates. Remove the backbones and discard. Cut the remaining fillets in half lengthwise.

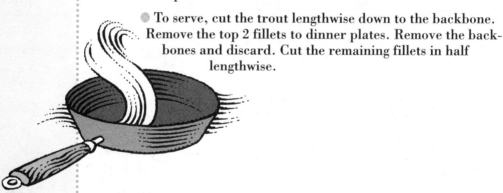

SALMON AND ONIONS BAKED IN FOIL

Baking fish in foil keeps it moist. You can assemble these packets with fresh salmon bought on sale, then freeze the packets for up to 1 month before baking. If making the packets to freeze, let the vegetable mixture chill before putting it over the raw fillets.

¼ cup white wine or nonalcoholic white wine
2 cups frozen sliced onions
1 cup frozen chopped green or sweet red peppers
3 cloves garlic, minced
½ teaspoon dried thyme
⅛ teaspoon dried sage
4 frozen salmon fillets (4 ounces each)
¼ teaspoon salt
¼ teaspoon ground black pepper

Makes 4 servings

Per serving
Calories 217
Total fat 7.5 g.
Saturated fat 1.2 g.
Cholesterol 64 mg.
Sodium 198 mg.
Fiber 2.5 g.

Cost per serving

$1.35

✸ Preheat the oven to 400°F.

✸ Coat a 10″ no-stick skillet with no-stick spray. Add the wine and bring to a boil over medium-high heat. Add the onions, green or red peppers, garlic, thyme, and sage. Cook, stirring, for 10 to 12 minutes, or until the onions are very soft.

✸ Place four 12″ × 12″ pieces of foil on a work surface. Place 1 salmon fillet on each piece of foil. Divide the vegetables evenly over the fillets. Sprinkle with the salt and pepper. Fold the sides of the foil over the fish to create sealed packets. Place the packets on a large baking sheet.

✸ Bake for 10 to 12 minutes, or until the fish is opaque in the center. Check by carefully opening 1 packet and inserting the tip of a sharp knife into 1 fillet.

Per serving
Calories 328
Total fat 10.1 g.
Saturated fat 2.5 g.
Cholesterol 147 mg.
Sodium 649 mg.
Fiber 2.3 g.

Cost per serving

65¢

SALMON CAKES WITH HORSERADISH CREAM

Assembled in 10 minutes or less, these cakes can be a thrifty lifesaver for people on busy schedules.

Salmon Cakes

2 cups frozen shredded potatoes
1 can (16 ounces) salmon, flaked
½ cup frozen chopped onions
2 eggs, lightly beaten
3 tablespoons nonfat mayonnaise
1 tablespoon reduced-sodium
      Worcestershire sauce
1 teaspoon lemon juice
½ teaspoon dried thyme
½ teaspoon ground black pepper
⅛ teaspoon paprika
¼ cup all-purpose flour

Horseradish Cream

3 tablespoons nonfat sour cream
1 teaspoon prepared horseradish
½ teaspoon honey
2 tablespoons minced fresh chives

✳ *To make the salmon cakes:* In a medium bowl, combine the potatoes, salmon, onions, eggs, mayonnaise, Worcestershire sauce, lemon juice, thyme, pepper, and paprika. Mix well and form into 4 cakes. Dredge in the flour. Place on a plate. Cover and refrigerate for 10 minutes.

✳ Coat a 10″ no-stick skillet with no-stick spray and place over medium-high heat until hot. Add the cakes. Cook for 3 minutes. Turn and cook for 3 minutes, or until firm and golden brown.

✳ *To make the horseradish cream:* In a small bowl, combine the sour cream, horseradish, and honey. Mix well. Spoon over the cakes. Sprinkle with the chives.

SALMON IN MUSTARD SAUCE

Freezing the uncooked salmon directly in the mustard sauce not only saves time on rushed evenings but also allows the acid in the wine to soften the fibers of the fish and make it plumper. This technique also works well for any ocean fish, such as mahi mahi, grouper, halibut, or whiting.

Makes 4 servings

Per serving
Calories 227
Total fat 11.1 g.
Saturated fat 1.6 g.
Cholesterol 64 mg.
Sodium 148 mg.
Fiber 0.3 g.

Cost per serving

$1.41

½ cup white wine or nonalcoholic white wine
2 tablespoons chopped onions
2 tablespoons minced garlic
1 tablespoon olive oil
1 tablespoon Dijon mustard
½ teaspoon ground paprika
4 salmon fillets (4 ounces each)

❋ In a shallow nonmetal baking dish, combine the wine, onions, garlic, oil, mustard, and paprika. Mix well. Add the salmon and turn to coat all sides. Cover and refrigerate for 1 hour, turning occasionally.

❋ Preheat the broiler. Coat the broiler pan with no-stick spray. Remove the fish from the marinade; reserve the marinade. Place the fish on the broiler pan. Broil 4″ from the heat, basting frequently with the reserved marinade, for 8 to 10 minutes, or until the fish is opaque in the center. Check by inserting the tip of a sharp knife into 1 fillet.

KITCHEN TIP

To freeze the uncooked marinated salmon, pack in a freezer-quality plastic bag. Seal, then wrap in freezer-quality foil. To use, thaw overnight in the refrigerator. Cook according to the recipe directions.

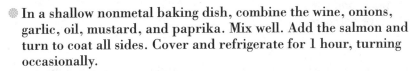

SAVE $2.50 IN FIVE MINUTES

Cut your own fish steaks from a whole fish such as a small salmon, pike, walleye, or whitefish. A 2-pound whole fish, cleaned and ready to fillet, will give you about six small steaks. The whole fish costs about $1.25 less a pound than ready-cut steaks.

Using a sharp knife, remove the skin in long strips, pulling it away from the fish. Then, following the natural breaks in the cartilage and bone, slice the fish into 4-ounce steaks. Trim and discard any dark areas of the flesh. Freeze the head and tail for a future Fish Stock (page 60).

SALMON SUPPER IN A PACKET

The orange marinade plays up the bright pink color of the fish, creating a pretty contrast with the spinach and peaches. None of the frozen vegetables requires thawing except the spinach, which takes just minutes in the microwave.

¼ cup nonfat plain yogurt
1 tablespoon frozen orange juice concentrate
2 cups packed frozen spinach
1 package (10 ounces) frozen spinach, thawed
¾ cup frozen sliced peaches
4 frozen salmon fillets (4 ounces each)
1 tablespoon reduced-sodium soy sauce
1 teaspoon lemon juice
1 clove garlic, minced

❋ Preheat the oven to 425°F.

❋ In a small bowl, combine the yogurt and orange juice concentrate. Mix well.

❋ Place four 12″ × 12″ pieces of foil on a work surface. Place the spinach in a colander. Rinse under hot running water for 30 seconds. Squeeze the spinach to remove excess moisture. Divide the spinach among the pieces of foil. Top with the peaches and salmon. Sprinkle with the soy sauce, lemon juice, and garlic. Top with the yogurt mixture. Fold the sides of the foil over the fish, creating 4 sealed packets. Place the packets on a large baking sheet.

❋ Bake for 10 to 12 minutes, or until the fish is opaque in the center. Check by carefully opening 1 packet and inserting the tip of a sharp knife into 1 fillet.

Snapper with Orange Sauce

Despite their expensive reputation, red snapper fillets are often on sale for as little as $4.99 a pound. With ingredients that you have right on your pantry shelf, you can whip up this exotic entrée from the freezer.

½   cup frozen defatted Fish Stock (page 60), thawed
½   cup reduced-calorie orange marmalade
 1   teaspoon dark sesame oil
½   teaspoon Dijon mustard
⅓   cup all-purpose flour
 1   teaspoon paprika
⅛   teaspoon salt
⅛   teaspoon ground black pepper
 4   frozen red snapper fillets (4 ounces each), thawed

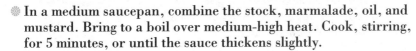

※ In a medium saucepan, combine the stock, marmalade, oil, and mustard. Bring to a boil over medium-high heat. Cook, stirring, for 5 minutes, or until the sauce thickens slightly.

※ In a shallow dish, combine the flour, paprika, salt, and pepper. Mix well.

※ Coat a 10″ no-stick skillet with no-stick spray. Place over medium-high heat until hot. Dip the snapper into the flour mixture and place in the skillet. Coat with no-stick spray. Cook for 5 minutes, or until golden brown. Turn and cook for 5 to 8 minutes, or until the fish is opaque. Check by inserting the tip of a sharp knife into the center of 1 fillet. Serve with the sauce.

Perfect Fish Every Time!

How can you tell if fish is cooked properly—not so much that it becomes tough and rubbery but enough so that it is safe to eat?

Chefs often judge fish cooking time by the thickness of the fillet or steak. Allow 8 to 10 minutes of cooking time per inch of fish if broiling, grilling, or baking at an oven temperature higher than 400°F. Fish cooked below 400°F requires another few minutes per inch.

Check for doneness by inserting the tip of a sharp knife in the fillet or steak to see if the interior is opaque yet still moist and tender.

Per serving
Calories 278
Total fat 3.3 g.
Saturated fat 0.7 g.
Cholesterol 37 mg.
Sodium 204 mg.
Fiber 5.5 g.

Cost per serving

$1.72

KITCHEN TIP

To freeze, pack the cooled cooked cod in a freezer-quality plastic container. To use, thaw overnight in the refrigerator. Cover and microwave on high power for 3 to 5 minutes, or until hot.

SOUTH-OF-FRANCE BAKED COD

The fragrance of tomatoes, capers, and olives in this sauce will bring back summer any time of the year. Make extra in August, when peppers are on sale or ripe from your garden, and freeze for fall and winter menus.

1½ cups chopped onions
1 green pepper, chopped
1 sweet red pepper, chopped
¼ cup apple juice
4 cups chopped tomatoes
2 cups cubed red potatoes
¼ cup chopped fresh parsley
1 teaspoon minced garlic
1 tablespoon capers, rinsed
2 teaspoons packed brown sugar
1 teaspoon chopped pitted black olives
4 cod fillets (4 ounces each)

❀ Preheat the oven to 400°F.

❀ Coat a 10″ no-stick skillet with no-stick spray and place over medium-high heat until hot. Add the onions, green peppers, red peppers, and apple juice. Cook, stirring, for 5 minutes, or until the peppers are soft. Add the tomatoes, potatoes, parsley, and garlic. Cook, stirring occasionally, for 10 minutes, or until the sauce thickens slightly. Add the capers, brown sugar, and olives. Stir well.

❀ Spread half of the sauce in the bottom of a 13″ × 9″ baking dish. Place the cod on top. Cover with the remaining sauce.

❀ Cover and bake for 15 minutes, or until the fish is opaque in the center. Check by inserting the tip of a sharp knife into the center of 1 fillet.

Tuna-Spaghetti Casserole

This twist on a traditional favorite is packed with tasty vegetables. It can be assembled in minutes, then frozen to save you time—and money—on a rushed night.

1 teaspoon olive oil
2 cups sliced mushrooms
3 cloves garlic, minced
1 tablespoon all-purpose flour
2 cups skim milk
½ cup shredded low-fat extra-sharp
   Cheddar cheese
8 ounces spaghetti
2 cups broccoli florets
2 cans (7 ounces each) water-packed tuna,
   drained and flaked
½ cup shredded carrots
¼ cup chopped fresh parsley
¼ cup dry bread crumbs
¼ teaspoon salt
¼ teaspoon ground black pepper

❋ Preheat the oven to 400°F. Coat a 13" × 9" freezer-proof baking dish with no-stick spray.

❋ Coat a 10" no-stick skillet with no-stick spray and place over medium-high heat until hot. Add the oil, mushrooms, and garlic. Cook, stirring occasionally, for 5 to 8 minutes, or until the mushrooms are very soft.

❋ Reduce the heat to low. Add the flour (the mixture will be dry). Cook, stirring, for 2 minutes. Gradually add the milk and Cheddar. Cook, stirring frequently, for 5 to 8 minutes, or until the sauce thickens. Remove from the heat.

❋ Cook the spaghetti in a large pot of boiling water according to the package directions. Drain well. Transfer to a large bowl. Add the broccoli, tuna, carrots, and mushroom sauce. Toss well. Place in the prepared baking dish.

❋ In a small bowl, combine the parsley, bread crumbs, salt, and pepper. Mix well. Sprinkle over the tuna mixture.

❋ Bake for 15 to 20 minutes, or until bubbly.

Makes 6 servings

Per serving
Calories 321
Total fat 4.3 g.
Saturated fat 1.7 g.
Cholesterol 23 mg.
Sodium 433 mg.
Fiber 4.3 g.

Cost per serving

61¢

Kitchen Tip

To freeze, cool the baked casserole. Wrap the entire baking dish in freezer-quality plastic wrap, then in freezer-quality foil. To use, remove the foil and thaw overnight in the refrigerator. Remove the plastic wrap. Cover and microwave on high power for 8 to 10 minutes, or until hot.

Per serving
Calories 305
Total fat 1.1 g.
Saturated fat 0.3 g.
Cholesterol 55 mg.
Sodium 255 mg.
Fiber 2.2 g.

Cost per serving

$1.60

KITCHEN TIP

To freeze, pack the cooled cooked rice in a freezer-quality plastic container. Top with the cooled cooked seafood. To use, thaw overnight in the refrigerator. Cover and microwave on high power for 5 minutes, or until hot.

CURRIED SEAFOOD AND VEGETABLES OVER RICE

Curry-spiced fish is sautéed then served over a bed of vegetables and rice. Make extra to freeze for a quick, satisfying entrée.

1	cup long-grain white rice
½	cup minced onions
½	cup minced sweet red peppers
½	cup shredded carrots
½	cup minced celery
2	scallions, chopped
8	ounces cod fillets, cut into 1" cubes
4	ounces medium shrimp, peeled and deveined
2	tablespoons white wine or apple juice
2	teaspoons reduced-sodium soy sauce
1	teaspoon minced garlic
¼	teaspoon curry powder
⅛	teaspoon salt

✻ Cook the rice according to the package directions.

✻ Coat a 10″ no-stick skillet with no-stick spray and place over medium-high heat until hot. Add the onions, peppers, carrots, celery, and scallions. Cook, stirring, for 2 minutes, or until the vegetables are bright in color. Transfer to a medium bowl.

✻ To the skillet, add the cod, shrimp, wine or apple juice, soy sauce, garlic, curry powder, and salt. Cover and cook, stirring frequently, for 4 to 6 minutes, or until the fish and shrimp are opaque in the center. Check by inserting the tip of a sharp knife into the center of 1 cube and 1 shrimp.

✻ Add the vegetables to the skillet. Toss to combine. Heat through. Serve over the rice.

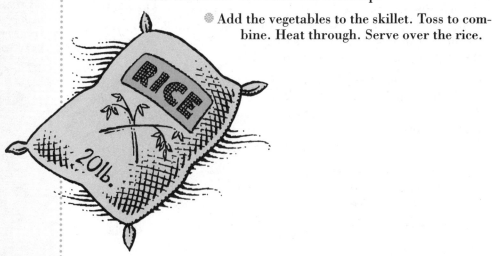

Seafood and Pasta Casserole

This fast and healthy meal saves you $6 a serving over restaurant versions. It can even be made with previously frozen fish because the sauce keeps the seafood moist and tender.

½ cup reduced-sodium vegetable juice
1 cup diced onions
1 green pepper, diced
2 cloves garlic, minced
1 can (16 ounces) Italian-style tomatoes
  (with juice)
1 cup frozen defatted Fish Stock (page 60),
  thawed
8 ounces spaghetti, broken
4 ounces medium shrimp, peeled and deveined
4 ounces cod fillets, cut into 1" cubes
½ teaspoon ground black pepper
¼ teaspoon salt

✻ Coat a Dutch oven with no-stick spray. Add the vegetable juice and bring to a boil over medium-high heat. Add the onions, green peppers, and garlic. Cook, stirring, for 5 minutes, or until the onions are soft but not browned.

✻ Add the tomatoes (with juice), stock, spaghetti, shrimp, and cod. Bring to a boil. Reduce the heat to medium. Cover and cook for 15 minutes, or until thick. Stir in the black pepper and salt.

Makes 4 servings

Per serving
Calories 329
Total fat 2.7 g.
Saturated fat 0.3 g.
Cholesterol 54 mg.
Sodium 531 mg.
Fiber 3.3 g.

Cost per serving

$1.23

Kitchen Tip

To freeze, pack the cooled cooked seafood in a freezer-quality plastic container. To use, thaw overnight in the refrigerator. Microwave on high power for 5 minutes, or until hot.

Per serving
Calories 222
Total fat 4.6 g.
Saturated fat 0.8 g.
Cholesterol 61 mg.
Sodium 459 mg.
Fiber 9.5 g.

Cost per serving

$1.74

KITCHEN TIP

*Grate a whole knob
of fresh ginger
and freeze in
1 teaspoon dollops
on a baking sheet
lined with wax
paper. Transfer to
a freezer-quality
plastic bag or
container and
freeze for up to a
year. When a recipe
calls for ginger,
just pull out the
amount that you
need. Freezing your
own ginger saves
45¢ per ½ cup over
jarred chopped
ginger.*

CHINESE SHRIMP AND VEGETABLES

*Even with the addition of cooked noodles or rice, this fast one-dish meal
costs less than $2 a serving.*

6   ounces frozen uncooked peeled medium
       shrimp
2   tablespoons reduced-sodium soy sauce
1   tablespoon dark sesame oil
1   teaspoon honey
1   teaspoon cornstarch
¼   cup frozen defatted Chicken Stock
       (page 61), thawed
2   cups frozen chopped onions
2   teaspoons minced garlic
1   teaspoon grated fresh ginger
2   cups frozen chopped sweet red or green peppers
1   package (10 ounces) frozen artichoke
       hearts, chopped
1   cup frozen whole kernel corn
1   cup frozen peas
¼   cup chopped fresh cilantro

In a medium bowl, combine the shrimp, soy sauce, oil, honey,
and cornstarch. Toss well. Cover and refrigerate for 3 hours,
stirring occasionally.

Coat a 10″ no-stick skillet with no-stick spray. Add the stock and
bring to a boil over medium-high heat. Add the onions, garlic,
and ginger. Cook, stirring, for 3 minutes, or until the onions
soften slightly.

Add the peppers, artichoke hearts, and corn. Cook, stirring, for
5 minutes, or until the peppers are soft. Add the shrimp and
marinade. Cook, stirring, for 5 minutes, or until the shrimp is
bright pink and opaque in the center. Check by inserting the tip
of a sharp knife into 1 shrimp.

Add the peas and cilantro. Cover and cook for 1 minute, or until
the peas are bright green.

Linguine with Clam Sauce

Make this Italian seafood dish in multiple batches and freeze for up to 2 months. If desired, you can make just the sauce to freeze and add freshly cooked linguine at serving time.

Clam Sauce

¼ cup apple juice
1 teaspoon olive oil
1 cup chopped onions
6 large cloves garlic, minced
1 cup dry white wine or frozen defatted
   Fish Stock (page 60), thawed
3 cans (6 ounces each) chopped clams (with juice)
1 cup chopped reduced-sodium canned tomatoes,
   drained
¼ teaspoon crushed red-pepper flakes

Linguine

1 pound linguine
¼ cup chopped fresh parsley
¼ cup grated Parmesan cheese
½ teaspoon ground black pepper

❋ **To make the clam sauce:** In a 10″ no-stick skillet over medium-high heat, bring the apple juice and oil to a boil. Add the onions and garlic. Cook, stirring, for 5 to 8 minutes, or until the onions are very soft but not browned. Add the wine or stock and bring to a boil. Cook, stirring occasionally, for 10 minutes, or until the liquid is reduced to ¼ cup.

❋ Add the clams (with juice), tomatoes, and red-pepper flakes. Cook for 2 minutes, or until hot.

❋ **To make the linguine:** Cook the linguine in a large pot of boiling water according to the package directions. Drain and toss with the clam sauce. Sprinkle with the parsley, Parmesan, and pepper.

Makes 6 servings

Per serving
Calories 450
Total fat 4.5 g.
Saturated fat 1.2 g.
Cholesterol 36 mg.
Sodium 241 mg.
Fiber 4.7 g.

Cost per serving

$1.00

Kitchen Tip

To freeze, pack the cooled cooked linguine and clam sauce in a freezer-quality plastic container. To use, thaw overnight in the refrigerator. Cover and microwave on high power for 3 to 5 minutes, or until hot.

Per kabob
Calories 130
Total fat 1.9 g.
Saturated fat 0.4 g.
Cholesterol 58 mg.
Sodium 488 mg.
Fiber 1.5 g.

Cost per serving

$1.99

KITCHEN TIP

For special
occasions, you can
double the
scallops at an
increase of only
50¢ a serving.

SEAFOOD KABOBS

Kabobs are the most economical way to get a variety of tasty seafood in one dish without breaking the bank. When you buy bulk frozen loose-pack scallops and shrimp, you can remove just the amount you need for your recipe without thawing the whole bag.

¼ cup apple juice
1 tablespoon hoisin sauce
1 tablespoon reduced-sodium soy sauce
1 teaspoon sugar
1 teaspoon cider vinegar
1 cup frozen broccoli florets
1 cup frozen pearl onions
1 small sweet red pepper, cut into 1" cubes
4 ounces frozen halibut steak, cut into 1" cubes
4 ounces frozen uncooked peeled medium shrimp
4 frozen scallops, quartered

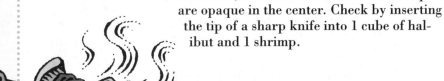

❋ In a small bowl, combine the apple juice, hoisin sauce, soy sauce, sugar, and vinegar. Mix well.

❋ Alternately thread the broccoli, onions, peppers, halibut, shrimp, and scallops onto four 12" metal skewers. Place the skewers in a 13" × 9" baking dish. Drizzle with the apple-juice mixture. Cover and refrigerate for 30 minutes, brushing frequently with the marinade.

❋ Preheat the broiler. Broil the kabobs 4" from the heat, basting frequently with the marinade, for 3 minutes, or until brown. Turn and broil for 3 minutes, or until the halibut and shrimp are opaque in the center. Check by inserting the tip of a sharp knife into 1 cube of halibut and 1 shrimp.

SEAFOOD SPINACH SALAD

To keep this salad economical as well as tasty, a small amount of scallops and shrimp is supplemented with plenty of seasonal fresh vegetables. It's a great entrée for hot summer days because you can take advantage of bargain buys on fresh corn. In the winter, frozen corn works well, too.

8   ounces frozen uncooked peeled shrimp
4   ounces frozen scallops, halved
½   cup frozen chopped onions
2   cloves garlic, minced
¼   cup frozen defatted Chicken Stock
      (page 61), thawed
4   cups chopped spinach
2   cups chopped tomatoes
1   cup frozen whole kernel corn
⅓   cup frozen chopped sweet red peppers
¼   cup crumbled feta cheese
3   tablespoons lemon juice
1   tablespoon olive oil

✤ Coat a 10″ no-stick skillet with no-stick spray and place over medium-high heat until hot. Add the shrimp, scallops, onions, and garlic. Cook, stirring, for 3 minutes, or until the seafood turns golden brown. Add the stock. Bring to a boil. Cook for 3 minutes, or until the seafood is opaque in the center. Check by inserting the tip of a sharp knife into 1 scallop or shrimp. Remove from the heat.

✤ In a large bowl, combine the spinach, tomatoes, corn, peppers, feta, lemon juice, and oil. Toss well. Add the seafood mixture. Toss well.

Makes 4 servings

Per serving
Calories 205
Total fat 8.1 g.
Saturated fat 3 g.
Cholesterol 99 mg.
Sodium 364 mg.
Fiber 3.5 g.

Cost per serving

$1.99

Per serving
Calories 353
Total fat 4.6 g.
Saturated fat 1 g.
Cholesterol 48 mg.
Sodium 327 mg.
Fiber 3.4 g.

Cost per serving

$1.20

KITCHEN TIP

To freeze, pack the
cooled cooked
shrimp and orzo in
a freezer-quality
plastic container.
To use, thaw
overnight in the
refrigerator. Cover
and microwave on
high power for
3 to 5 minutes,
or until hot.

SHRIMP SKILLET SUPPER WITH ORZO

*Look for orzo, a rice-shaped pasta, in your supermarket. The sauce makes
this dish an excellent candidate for freezing.*

1½ cups orzo pasta
 1 teaspoon olive oil
 1 cup minced onions
 2 cloves garlic, minced
 8 ounces medium shrimp, peeled, deveined,
    and halved lengthwise
   Grated rind and juice of 1 lemon
¾ cup white wine or frozen defatted Fish Stock
    (page 60), thawed
¼ teaspoon salt
¼ teaspoon ground black pepper
¼ cup minced fresh parsley
 2 tablespoons grated Parmesan cheese

❀ Cook the pasta in a large pot of boiling water according to the
package directions. Drain well.

❀ Coat a 10″ no-stick skillet with no-stick spray and place over
medium-high heat until hot. Add the oil, onions, and garlic.
Cook, stirring, for 5 minutes, or until the onions are soft but not
browned. Add the shrimp. Cook for 3 minutes. Turn and cook
for 3 minutes, or until the shrimp is opaque in the center. Check
by inserting the tip of a sharp knife into 1 shrimp.

❀ Add the lemon rind, lemon juice, wine or stock, salt, and
pepper. Bring to a boil. Reduce the heat to medium and cook for
2 to 3 minutes, or until the liquid is reduced by half.

❀ Add the pasta, parsley, and Parmesan. Toss well.

THE SKINNY ON SHRIMP

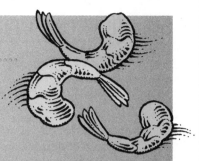

Americans don't skimp on shrimp. In fact, we consume more than 800 million pounds of shrimp a year, making it the most popular fresh/frozen seafood in the country.

Virtually all shrimp is frozen after harvesting so for better quality when stocking your freezer, select bulk frozen shrimp instead of the thawed (previously frozen) shrimp at the fish counter. Purchase loose-pack bagged frozen shrimp in the midsize range, with about 31 to 40 shrimp in each pound. Slightly larger and slightly smaller shrimp will also work for the recipes in this book. Just increase or decrease the cooking time by about 1 minute more or less. Shrimp are cooked when the flesh is opaque all the way through. Check for doneness by cutting into a shrimp. Take care not to overcook shrimp or they will toughen.

You can keep shrimp in the freezer for up to three months without loss of flavor. Thaw frozen shrimp overnight in the refrigerator or thaw in the microwave just before cooking. If you forget to thaw in advance, just place the shrimp in a fine strainer and submerge in a saucepan of boiling water for 1 to 2 minutes, or until pliable. Or, if you're making a soup, stew, or other saucy dish, just add the frozen shrimp directly to the recipe and add a few minutes to the allotted cooking time.

Shrimp with Puttanesca Sauce

Shrimp pairs perfectly with this robust sauce, a favorite in Italian cooking. Spaghetti and lots of vegetables in the dish keep the cost per serving reasonable.

2 teaspoons olive oil
2 cups frozen chopped onions, thawed
2 teaspoons minced garlic
1 can (28 ounces) reduced-sodium whole tomatoes, chopped (with juice)
1 cup chopped sweet red or green peppers
¼ cup white wine or apple juice
1 teaspoon capers, rinsed
½ teaspoon Worcestershire sauce
¼ teaspoon crushed dried rosemary
1 drop hot-pepper sauce
12 ounces medium shrimp, peeled and deveined
8 ounces spaghetti
2 tablespoons chopped fresh parsley

✤ Coat a 10″ no-stick skillet with no-stick spray and place over medium-high heat until hot. Add the oil, onions, and garlic. Cook, stirring, for 5 minutes, or until the onions are soft but not browned. Add the tomatoes (with juice), peppers, wine or apple juice, capers, Worcestershire sauce, rosemary, and hot-pepper sauce. Cook for 8 minutes, or until the sauce is thick.

✤ Add the shrimp. Cook for 5 minutes, or until the shrimp is opaque in the center. Check by inserting the tip of a sharp knife into 1 shrimp.

✤ Cook the spaghetti in a large pot of boiling water according to the package directions. Drain well. Toss with the sauce. Sprinkle with the parsley.

Makes 4 servings

Per serving
Calories 397
Total fat 4.8 g.
Saturated fat 0.8 g.
Cholesterol 122 mg.
Sodium 245 mg.
Fiber 8.2 g.

Cost per serving

$1.54

Kitchen Tip

To freeze, pack the cooled cooked shrimp, spaghetti, and sauce in a freezer-quality plastic container. To use, thaw overnight in the refrigerator. Cover and microwave on high power for 5 minutes, or until hot.

FAST FISH DINNERS

What's for dinner? If you have fish in the freezer, the question is answered. Thaw fish fillets overnight in the refrigerator, then prepare with one of these speedy main-dish methods.

- *Steamed Fish*. Place cod or haddock fillets in a shallow baking dish; add 1 inch of fish stock or defatted chicken stock to the dish. Sprinkle with chopped fresh herbs, if desired. Cover and bake at 350°F for 12 minutes, or until the fish is opaque in the center. Test by inserting the tip of a sharp knife in the center of a fillet.

- *Grilled Fish*. Spray the grill rack with no-stick spray. Brush salmon, red snapper, or orange roughy fillets with a small amount of olive oil and sprinkle with ground black pepper. Grill 4 inches from the heat for 5 minutes, then turn and grill for 3 minutes, or until the fish is opaque in the center. Test by inserting the tip of a sharp knife in the center of a fillet.

- *Fish Kabobs with Vegetables*. Alternate cubes of salmon, tuna, or other firm-fleshed fish with pineapple chunks, frozen pearl onions, and wedges of green peppers. Coat the kabobs with no-stick spray. Grill or broil 4 inches from the heat, turning occasionally, for 5 minutes, or until the fish is opaque in the center. Test by inserting the tip of a sharp knife in the center of a fillet.

CHAPTER 7

PASTA, VEGETABLE, AND EGG MAIN DISHES

••

Whole grains, beans, and pasta shape so many meals that meatless is fast becoming mainstream. Each bite gives you good eating in the form of complex carbohydrates, fiber, vitamins, and minerals with virtually no fat. And you spend only 17 to 75 cents per pound for meatless proteins—a real bargain compared with lean meat, poultry, or fish.

Even if your lifestyle is fast-paced, grains and beans fit in fine—thanks to the freezer. Cooking two or three times the amount of rice or beans is no extra work, but it gives you a stockpile that you can freeze for up to six months without loss of flavor or texture. Fast and economical meatless meals become a snap to prepare with a well-stocked freezer.

If you prefer to cook dishes for the freezer, most cooked meatless main dishes—casseroles, burritos, and ragoûts—freeze very well.

Per serving
Calories 385
Total fat 5.1 g.
Saturated fat 2.6 g.
Cholesterol 15 mg.
Sodium 456 mg.
Fiber 9.8 g.

Cost per serving

82¢

LASAGNA BLACK BEAN ROLL-UPS

Layering cooked black beans with the vegetables adds 5 grams of fiber per serving. Rolling the lasagna noodles makes this much prettier than the typical pasta casserole. If you freeze your cooked beans in 1-cup packages, they'll thaw quickly in the microwave.

12 lasagna noodles
 1 cup frozen chopped onions
 2 tablespoons water
 1 tablespoon frozen apple juice concentrate
 2 cloves garlic, minced
 2 cups thinly sliced mushrooms
 2 cups frozen cooked black beans (page 177), thawed
 1 cup frozen sliced carrots, chopped
 1 teaspoon chili powder
½ teaspoon salt
½ teaspoon ground black pepper
1½ cups low-fat ricotta cheese
¼ cup grated Parmesan cheese
 2 cans (14 ounces each) reduced-sodium tomato sauce
¼ cup shredded low-fat extra-sharp Cheddar cheese

❋ Preheat the oven to 350°F. Coat a 13″ × 9″ baking dish with no-stick spray.

❋ Cook the noodles in a large pot of boiling water according to the package directions. Drain well. Rinse under cold water. Place on a tray in a single layer.

❋ In a 10″ no-stick skillet, combine the onions, water, apple juice concentrate, and garlic. Cook, stirring, over medium-high heat for 5 minutes, or until the onions soften. Add the mushrooms, beans, carrots, chili powder, salt, and pepper. Cook, stirring, for 4 minutes. Remove from the heat. Mash the beans and vegetables with a fork.

❋ In a blender or food processor, combine the ricotta and Parmesan. Process until smooth. Stir into the vegetable mixture.

❋ Spread a generous amount of the vegetable mixture over the length of each noodle, then roll up. Place the rolls, seam side down, in the prepared baking dish. Top with the tomato sauce. Sprinkle with the Cheddar. Cover and bake for 25 minutes, or until bubbly.

FOUR-CHEESE MACARONI BAKE

Makes 8 servings

Per serving
Calories 301
Total fat 4.7 g.
Saturated fat 2 g.
Cholesterol 13 mg.
Sodium 438 mg.
Fiber 2.1 g.

Cost per serving

33¢

At only 33¢ a serving, this creamy casserole is a freezer success. You can lower the price another 10¢ a serving if you buy macaroni on sale for 79¢ a pound.

3 cups elbow macaroni
1 cup minced onions
2 cloves garlic, minced
1 teaspoon olive oil
⅓ cup all-purpose flour
3 cups skim milk
¾ cup nonfat cottage cheese
½ cup shredded low-fat extra-sharp Cheddar cheese
½ cup shredded nonfat mozzarella cheese
½ cup grated Parmesan cheese
¼ teaspoon salt
¼ teaspoon ground black pepper
½ cup dry bread crumbs

❋ Preheat the oven to 375°F. Coat a 13" × 9" freezer-proof baking dish with no-stick spray.

❋ Cook the macaroni in a large pot of boiling water according to the package directions. Drain well. Place in a large bowl.

❋ Coat a large saucepan with no-stick spray and place over medium-high heat until hot. Add the onions, garlic, and oil. Cook, stirring, for 5 minutes, or until the onions are golden brown. Add the flour. Cook, stirring, for 2 minutes (the mixture will be dry).

❋ In a blender or food processor, combine the milk and cottage cheese. Process until smooth. Add to the skillet. Cook, stirring constantly, for 5 to 7 minutes, or until the sauce thickens. Add the Cheddar, mozzarella, and ¼ cup of the Parmesan. Cook for 2 minutes, or until the cheeses melt.

❋ Stir in the macaroni, salt, and pepper. Spoon into the prepared baking dish.

❋ In a small bowl, combine the bread crumbs and the remaining ¼ cup Parmesan. Sprinkle over the macaroni. Bake for 30 minutes, or until the topping is golden brown.

KITCHEN TIP

To freeze, cool the cooked casserole. Wrap the baking dish in freezer-quality plastic wrap, then in freezer-quality foil. To use, thaw overnight in the refrigerator. Remove the foil and plastic wrap; discard the plastic wrap. Cover with the foil and bake at 350°F for 15 to 20 minutes, or until hot.

Pasta, Vegetable, and Egg Main Dishes

ROTELLE WITH HERBS AND CHEESE

Baked pasta with plenty of vegetables is an economical and satisfying main dish for only 97¢ a serving. Even a small amount of pungent herbs and black olives lends a uniquely Mediterranean flavor.

Per serving
Calories 439
Total fat 9.6 g.
Saturated fat 4.2 g.
Cholesterol 18 mg.
Sodium 404 mg.
Fiber 11 g.

Cost per serving

97¢

KITCHEN TIP

Get full value from any bunch of herbs you purchase. After using the amount you need for a recipe, chop and freeze the remaining leaves in a small freezer-quality plastic bag. Parsley and other herbs will remain green and flavorful. Use the herbs while still frozen.

8	ounces rotelle pasta
½	cup frozen defatted Chicken Stock (page 61), thawed
½	cup frozen chopped onions
½	cup frozen chopped sweet red peppers
½	cup chopped mushrooms
½	cup frozen broccoli florets
4	large cloves garlic, chopped
1	teaspoon olive oil
2	cans (28 ounces each) reduced-sodium tomatoes, chopped (with juice)
2	tablespoons tomato paste
1	tablespoon chopped black olives
1	teaspoon sugar
2	tablespoons chopped fresh parsley
1	teaspoon dried Italian herb seasoning
½	cup shredded low-fat mozzarella cheese
½	cup shredded low-fat Swiss cheese
¼	cup grated Parmesan cheese

Preheat the oven to 350°F. Coat a 13″ × 9″ baking dish with no-stick spray.

Cook the pasta in a large pot of boiling water according to the package directions. Drain well. Transfer to the prepared baking dish.

In a 10″ no-stick skillet, combine the stock, onions, peppers, mushrooms, broccoli, garlic, and oil. Cook, stirring, over medium-high heat for 5 minutes, or until the onions soften. Add the tomatoes (with juice), tomato paste, olives, and sugar. Cook, stirring, for 5 minutes, or until the tomatoes soften. Add the parsley and Italian herb seasoning.

Pour over the pasta. Sprinkle with the mozzarella, Swiss, and Parmesan.

Bake for 15 minutes, or until the sauce is thick and the cheese has melted.

MAKE ENDS MEATLESS

Frugality experts like Amy Dacyczyn, author of *The Tightwad Gazette* series, recommend extending your menus with meatless dishes to save plenty on your food budget. To make meatless meals appealing to your family and friends, use these handy tips from vegetarian chefs.

● Mix colors that appeal to the eye. For example, always enliven green vegetables with red, orange, or yellow (for example, broccoli with sweet red peppers or spinach with yellow summer squash).

● Build your entrées around low-cost protein sources like beans, whole grains, and tofu. Avoid the costly (and fat-laden) trap of too much cheese or cream sauces.

● Learn about herbs and spices, the keys to great flavor in any dish. Experiment with one new seasoning each week.

PENNE WITH SUMMER VEGETABLES

Use the ripest produce from your garden or your local farmers' market for this colorful pasta dish. If you like, you can package individual servings in freezer-quality plastic bags, then label and freeze for quick single-serving dinners.

¼ cup balsamic vinegar
2 tablespoons minced garlic
1 cup chopped onions
1 cup diced eggplant
1 cup diced yellow summer squash
1 sweet red pepper, diced
2 cups diced tomatoes
¼ cup chopped fresh parsley
1 tablespoon olive oil
¼ cup grated Parmesan cheese
1 teaspoon sugar
½ teaspoon grated orange rind
½ teaspoon salt
½ teaspoon chopped fresh thyme
8 ounces penne pasta

❉ In a 10" no-stick skillet, combine the vinegar and garlic. Cook, stirring, over medium-high heat for 3 minutes, or until the garlic softens. Add the onions, eggplant, squash, and peppers. Cook, stirring, for 5 minutes, or until the onions soften. Add the tomatoes and parsley. Cook, stirring frequently, for 15 minutes, or until the tomatoes are very soft. Add the oil, Parmesan, sugar, orange rind, salt, and thyme. Cook for 3 minutes, or until the sauce thickens.

❉ Cook the pasta in a large pot of boiling water according to the package directions. Drain well. Return to the pot. Add the sauce and toss well.

Vegetable Stroganoff over Noodles

When you have a craving for creamy pasta, trim your waistline and your budget with this satisfying stroganoff, made with plenty of vegetables and a low-fat sauce at less than $1 a serving.

12 ounces no-yolk noodles
¼ cup frozen defatted Chicken Stock (page 61), thawed
1 teaspoon olive oil
3 cups sliced mushrooms
1 cup frozen sliced carrots
2 cups frozen broccoli florets
2 cloves garlic, minced
¼ cup frozen chopped onions
¼ teaspoon caraway seeds, crushed
2 teaspoons cornstarch
¼ cup water
1 tablespoon tomato paste
1 cup low-fat sour cream
½ cup nonfat plain yogurt
1 tablespoon all-purpose flour
¼ teaspoon dried dillweed
¼ teaspoon salt
¼ teaspoon ground black pepper

✻ Cook the noodles in a large pot of boiling water according to the package directions. Drain well and place in a large bowl.

✻ In a 10" no-stick skillet, combine the stock and oil. Bring to a boil over medium-high heat. Add the mushrooms, carrots, broccoli, and garlic. Cook, stirring, for 5 minutes, or until the carrots soften slightly. Add the onions and caraway seeds. Cover and cook for 3 minutes.

✻ Place the cornstarch in a small bowl. Add the water and stir until smooth. Stir in the tomato paste. Add to the skillet. Cook, stirring, for 2 minutes, or until the sauce is thick.

✻ In a medium bowl, combine the sour cream, yogurt, flour, dillweed, salt, and pepper. Mix well. Add to the skillet, stirring well to combine. Cook for 1 minute, or until hot. Toss with the noodles.

Makes 4 servings

Per serving
Calories 480
Total fat 6.4 g.
Saturated fat 3.3 g.
Cholesterol 21 mg.
Sodium 331 mg.
Fiber 8.6 g.

Cost per serving

92¢

BAKED VEGETABLE PACKETS

Frozen vegetables can be baked in foil packets without thawing because the water generated as they bake steams them.

Per serving
Calories 229
Total fat 3.4 g.
Saturated fat 2 g.
Cholesterol 8 mg.
Sodium 223 mg.
Fiber 6.3 g.

Cost per serving

69¢

4 large baking potatoes, thinly sliced
1 cup frozen chopped onions
1 cup frozen cauliflower florets
1 cup frozen sliced carrots
¼ cup grated Parmesan cheese
2 tablespoons crumbled blue cheese
2 teaspoons minced garlic
¼ teaspoon ground black pepper

FROM the Freezer

❋ Preheat the oven to 450°F. Fold four 12″ × 12″ pieces of foil in half and place them on a large baking sheet.

❋ In a medium bowl, combine the potatoes, onions, cauliflower, carrots, Parmesan, blue cheese, garlic, and pepper. Toss well to combine. Divide among the pieces of foil. Fold the foil over the vegetables; tightly seal the edges to form 4 packets.

❋ Bake for 30 minutes, or until the vegetables are soft and the cheese is melted.

CHEESY BEAN BURRITOS

Pinto beans are loaded with fiber, adding about 5 grams a serving for just 16¢. If you freeze your cooked beans in 1-cup packages, they'll thaw quickly in the microwave. Remove the tortillas from the freezer before you start cooking and they will be thawed by the time the filling is ready.

 1 teaspoon olive oil
 ⅔ cup frozen chopped onions
 3 cloves garlic, minced
 1 cup frozen sliced carrots, chopped
 1 cup frozen whole kernel corn
 ½ cup frozen chopped sweet red or green peppers
 2 cups frozen cooked pinto beans (page 177), thawed
 ½ cup diced tomatoes
 4 frozen flour tortillas (10" diameter)
 ½ cup shredded nonfat Monterey Jack cheese
 ½ cup low-fat sour cream

❋ Preheat the oven to 350°F. Coat a 13" × 9" baking dish with no-stick spray.

❋ Coat a 10" no-stick skillet with no-stick spray and place over medium-high heat until hot. Add the oil, onions, and garlic. Cook, stirring, for 2 minutes. Add the carrots, corn, and peppers. Cook, stirring, for 3 minutes, or until the vegetables soften slightly.

❋ Add the beans and tomatoes. Mash the beans slightly with a fork. Cook, stirring frequently, for 5 minutes, or until the tomatoes soften.

❋ Warm the tortillas in the oven for 1 to 2 minutes, or until soft. Divide the bean mixture among the tortillas and roll them up to enclose the filling. Place, seam side down, in the prepared baking dish. Top with the Monterey Jack and sour cream. Cover and bake for 10 minutes, or until the cheese melts.

Makes 4 servings

Per serving
Calories 434
Total fat 7.6 g.
Saturated fat 2.9 g.
Cholesterol 11 mg.
Sodium 433 mg.
Fiber 11.4 g.

Cost per serving

72¢

Pasta, Vegetable, and Egg Main Dishes

ITALIAN STUFFED PEPPERS

Makes 4 servings

Per serving
Calories 207
Total fat 6.9 g.
Saturated fat 4.6 g.
Cholesterol 25 mg.
Sodium 427 mg.
Fiber 2 g.

Cost per serving

90¢

KITCHEN TIP

To freeze, pack the cooled cooked peppers in a single layer in a freezer-quality plastic container. To use, thaw overnight in the refrigerator. Microwave on high power for 7 to 8 minutes, or until hot.

Serve this summery main dish when sweet peppers are less than 25¢ each at the farmers' market—or ripe in your garden. Stuffed peppers taste good with a leafy salad, sliced tomatoes, and whole-grain bread.

4 large sweet red peppers
1 cup soft bread crumbs
1 cup shredded low-fat extra-sharp Cheddar cheese
½ cup chopped onions
½ cup frozen peas
½ cup shredded nonfat mozzarella cheese
¼ cup low-fat ricotta cheese
2 tablespoons frozen defatted Chicken Stock
    (page 61), thawed
4 cloves garlic, minced
1 tablespoon minced fresh basil
2 tablespoons grated Parmesan cheese

❋ Preheat the oven to 350°F. Slice the tops off the peppers and set aside. Remove the seeds and membranes. Place the peppers, cut side up, in a 8″ × 8″ baking dish.

❋ In a medium bowl, combine the bread crumbs, Cheddar, onions, peas, mozzarella, ricotta, stock, garlic, and basil. Mix well. Spoon into the peppers. Top with the Parmesan.

❋ Cover with the pepper tops and bake for 20 minutes, or until the peppers are soft. Remove the pepper tops and discard. Bake for 10 minutes, or until the topping is golden brown.

CHICKPEA SKILLET SUPPER

Chickpeas add 6 grams of fiber to each serving of this curried skillet supper. They freeze beautifully, so you can cook up a large batch (for only 21¢ a cup cooked) and have them on hand in the freezer in 1-cup packages that will thaw quickly in the microwave.

1 teaspoon olive oil
2 cups frozen chopped onions
1½ cups frozen sliced carrots
1 cup frozen broccoli florets
1 cup frozen cauliflower florets
3 large cloves garlic, chopped
3 cups frozen cooked chickpeas
   (page 177), thawed
1 cup frozen cooked navy beans
   (page 177), thawed
¾ cup apple juice
¾ teaspoon curry powder
½ teaspoon celery seeds
1 cup long-grain white rice
⅓ cup grated Parmesan cheese

❋ Coat a 10″ no-stick skillet with no-stick spray and place over medium-high heat until hot. Add the oil, onions, carrots, broccoli, cauliflower, and garlic. Cook, stirring, for 3 minutes, or until the broccoli is bright green. Add the chickpeas, beans, apple juice, curry powder, and celery seeds. Bring to a boil. Cover and cook for 35 minutes, or until thick.

❋ Cook the rice according to the package directions. Add to the skillet. Top with the Parmesan.

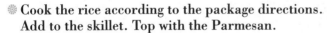

Per serving
Calories 597
Total fat 8 g.
Saturated fat 2.3 g.
Cholesterol 7 mg.
Sodium 230 mg.
Fiber 16.8 g.

Cost per serving

63¢

Pasta, Vegetable, and Egg Main Dishes

Per serving
Calories 271
Total fat 6.9 g.
Saturated fat 1.9 g.
Cholesterol 8 mg.
Sodium 166 mg.
Fiber 2.9 g.

Cost per serving

28¢

KITCHEN TIP

To freeze the cooled baked pizzas, wrap them in freezer-quality plastic wrap, then in freezer-quality foil. To use, thaw overnight in the refrigerator. Remove the foil and plastic wrap. Bake at 400°F for 15 minutes, or until hot.

PESTO-TOMATO PIZZA

Long after summer has turned to fall, you can savor these garden-fresh thin-crust pizzas from your freezer.

Dough

    1 cup warm water (about 110°F)
    1 tablespoon or 1 package active dry yeast
    1 tablespoon sugar
    ½ cup yellow cornmeal
    ¼ teaspoon salt
    2 tablespoons oil
    2½ cups all-purpose flour

Topping

    1 cup chopped fresh basil
    ½ cup soft bread crumbs
    2 cloves garlic, halved
    4 medium tomatoes, sliced
    1 cup shredded low-fat mozzarella cheese

● *To make the dough:* In a large bowl, combine the water, yeast, and sugar. Stir well. Let stand in a warm place for 5 minutes, or until foamy. Stir in the cornmeal, salt, and 1 tablespoon of the oil. Stir in up to 2¼ cups of the flour to make a kneadable dough.

● Turn the dough out onto a lightly floured surface. Knead, adding more of the remaining ¼ cup flour as necessary, for 10 minutes, or until smooth and elastic. Coat a large bowl with no-stick spray. Add the dough and turn to coat all sides. Cover loosely with a kitchen towel and set in a warm place for 15 minutes.

● Preheat the oven to 450°F. Coat two 12″ round pizza pans with no-stick spray. Divide the dough in half. Roll each half into a 12″ circle. Place the circles on the pizza pans. Brush the surface of each pizza with the remaining 1 tablespoon oil.

● *To make the topping:* In a blender or food processor, combine the basil, bread crumbs, and garlic. Process until smooth. Spread over the dough. Top with the tomatoes and mozzarella.

● Bake for 20 minutes, or until the crusts are golden brown and the cheese is bubbling.

Pizza Pronto!

Stock up on homemade thick-crust pizza shells when you have extra time. Shape the dough into circles and place on baking sheets. Place in the freezer until frozen. Then stack the shells between layers of freezer-quality plastic wrap and wrap tightly with freezer-quality foil. Freeze for up to four months.

You needn't thaw the shells before using them. Just add whatever toppings you desire. Bake at 450°F for 30 minutes (or add 10 minutes baking time to whatever recipe you're following).

Crusts made with this recipe cost you less than 80¢ each compared with $1.64 for frozen shells and $2.49 for seasoned baked pizza shells from the supermarket.

Thick-Crust Pizza Shells

- 2 cups warm water (about 110°F)
- 2 tablespoons or 2 packages active dry yeast
- 2 tablespoons sugar
- 1 cup yellow cornmeal
- ½ teaspoon salt
- 1 tablespoon oil
- 5 cups all-purpose flour

In a large bowl, combine the water, yeast, and sugar. Stir well. Let stand in a warm place for 5 minutes, or until foamy. Stir in the cornmeal, salt, and oil. Stir in enough of the flour to make a kneadable dough.

Turn the dough out onto a lightly floured surface. Knead, adding more flour as necessary, for 10 minutes, or until smooth and elastic. Coat a large bowl with no-stick spray. Add the dough and turn to coat all sides. Cover and set in a warm place for 15 minutes.

Divide the dough in half. Roll each half into a 12″ circle.

Makes 2 thick-crust shells; 8 servings per shell

Per serving
Calories 178
Total fat 0.7 g.
Saturated fat 0.1 g.
Cholesterol 0 mg.
Sodium 71 mg.
Fiber 2.1 g.

Cost per shell
76¢

Pasta, Vegetable, and Egg Main Dishes

PIZZA PRIMAVERA

Make these easy springtime pizzas with pita bread rounds, and your preparation will take just minutes. Feel free to use whatever frozen vegetables you have on hand.

- 2 whole-wheat pitas (6" diameter)
- 1 teaspoon oil
- 1 cup reduced-sodium pizza sauce
- ¼ cup frozen broccoli florets
- ¼ cup frozen artichoke hearts, chopped
- ¼ cup frozen chopped sweet red peppers
- ¼ cup frozen chopped onions
- 1 teaspoon dried Italian herb seasoning
- 1 cup shredded low-fat mozzarella cheese

❋ Preheat the oven to 400°F. Split the pitas into 2 thin disks each. Place, rough side up, on a large baking sheet. Brush the tops with the oil. Top with the pizza sauce. Sprinkle with the broccoli, artichoke hearts, peppers, and onions. Top with the Italian herb seasoning and mozzarella.

❋ Bake for 15 minutes, or until the cheese is bubbling.

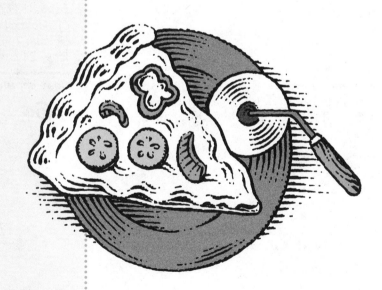

Makes 4

Per pizza
Calories 183
Total fat 6.8 g.
Saturated fat 3.3 g.
Cholesterol 15 mg.
Sodium 296 mg.
Fiber 4 g.

Cost per serving

37¢

CINNAMON-CHEESE BLINTZES

Makes 10

Per blintz
Calories 132
Total fat 3.5 g.
Saturated fat 1.7 g.
Cholesterol 74 mg.
Sodium 239 mg.
Fiber 0.3 g.

Cost per serving

22¢

Comfort wrapped in thin pancakes, blintzes are delightful for brunch or a light supper.

½ cup all-purpose flour
½ cup skim milk
3 eggs
1 teaspoon honey
½ teaspoon baking powder
2 cups nonfat cottage cheese
1 cup shredded low-fat mozzarella cheese
¼ cup packed brown sugar
1 teaspoon ground cinnamon

✻ In a blender or food processor, combine the flour, milk, eggs, honey, and baking powder. Process until very smooth.

✻ Coat a 10″ no-stick skillet with no-stick spray and place over medium-high heat until hot. Pour in ¼ cup of the batter and swirl the skillet to evenly coat the bottom. Cook for 1 minute, or until the top is set and the bottom is golden brown. Transfer to a plate. Repeat with the remaining batter to make a total of 10 blintzes.

✻ Preheat the oven to 300°F. Line a large baking sheet with foil.

✻ In the blender or food processor, combine the cottage cheese, mozzarella, brown sugar, and cinnamon. Process until smooth. Place 2 to 3 tablespoons of filling in the center of each blintz. Fold over the sides and ends to enclose the filling and form a packet. Place the blintzes, seam side down, on the prepared baking sheet. Bake for 10 minutes, or until hot.

KITCHEN TIP

To freeze, place the filled but unbaked blintzes on a tray. Place in the freezer until frozen. Pack the blintzes, separated by pieces of wax paper, in a freezer-quality plastic bag. To use, thaw overnight in the refrigerator. Place on a baking sheet. Cover loosely with foil. Bake at 300°F for 20 to 25 minutes, or until heated through.

Per serving
Calories 532
Total fat 3.8 g.
Saturated fat 0.5 g.
Cholesterol 2 mg.
Sodium 446 mg.
Fiber 17.6 g.

Cost per serving

95¢

KITCHEN TIP

To freeze, cool the
cooked bean pie in
the baking dish.
Wrap the entire
dish in freezer-
quality plastic
wrap, then in
freezer-quality foil.
To use, thaw
overnight in the
refrigerator.
Remove the foil.
Cover and
microwave on
high power for
5 minutes, or
until hot.

CRUSTY TAMALE-BEAN PIE

Cornmeal crust and a filling of vegetable chili make this pie hearty yet economical at only 95¢ a serving.

Crust

2 cups yellow cornmeal
¼ cup all-purpose flour
½ teaspoon baking powder
1 cup frozen defatted Chicken Stock
   (page 61), thawed

Filling

1 teaspoon olive oil
1 cup chopped onions
¼ cup dry sherry or apple juice
1 cup sliced carrots
1 cup whole kernel corn
1 cup chopped tomatoes
4 large cloves garlic, minced
2 cups frozen cooked kidney beans
   (page 177), thawed
¼ cup tomato paste
2 teaspoons ground cumin
2 teaspoons chili powder
½ cup canned diced green chili peppers
2 tablespoons frozen orange juice concentrate, thawed

❋ *To make the crust:* In a medium bowl, combine the cornmeal, flour, and baking powder. Mix well. Add the stock. Mix well, then knead for 2 minutes, or until smooth. Divide in half.

❋ Press half of the dough over the bottom and sides of an 8″ × 8″ no-stick baking dish. Wrap the remaining dough in plastic wrap. Let stand at room temperature while you make the filling.

❋ *To make the filling:* Coat a 10″ no-stick skillet with no-stick spray and place over medium-high heat until hot. Add the oil, onions, and sherry or apple juice. Cook, stirring, for 5 minutes, or until the onions are soft but not browned. Add the carrots, corn, tomatoes, and garlic. Cook, stirring, for 5 minutes, or until the carrots soften.

✻ Add the beans, tomato paste, cumin, and chili powder. Cook, stirring frequently, for 5 minutes, or until thick. Stir in the chili peppers. Spoon into the prepared baking dish.

✻ On a sheet of wax paper, pat the remaining dough into an 8″ × 8″ square. Invert over the filling and remove the wax paper. Brush the crust with the orange juice concentrate. Bake for 25 minutes, or until the crust is golden brown and the filling is bubbly.

CHILI PEPPERS ADD PIZZAZZ

Explore the wide range of chili peppers available—fresh, jarred, canned, and dried— to ignite flavors in low-fat cooking. You'll find a wide variety of inexpensive chili peppers at Asian and Hispanic markets and many large supermarkets. Follow this rule of thumb for fresh and dried chili peppers: The longer and greener fresh chili peppers are best for stuffing; the short, red fresh or dried ones are a zestful addition to soups, stews, and other saucy dishes.

TOFU TEXTURIZER

Meat takes the biggest bite out of your food budget. But you can bite back. Go meatless for just a few meals a week and see the savings add up. Tofu, at 75¢ a pound, is a bargain and supplies protein in lasagna, spaghetti sauces, and chili.

To mimic the texture of ground meat, all you have to do is freeze tofu before using. An added bonus to previously frozen tofu is that it absorbs flavoring much better than unfrozen tofu.

To freeze tofu, unwrap and drain, then place on a freezer-proof plate or baking sheet. Freeze until solid, then package in freezer-quality plastic bags. Store for up to six months. Thaw overnight in the refrigerator, then squeeze out all liquid. Crumble and add to any recipe in place of ground beef or turkey.

LAYERED BLACK BEAN CASSEROLE

Black beans and low-fat tofu replace the traditional ground beef in this satisfying enchilada casserole.

2   cups frozen cooked black beans
     (page 177), thawed
4   ounces reduced-fat extra-firm tofu
½   cup salsa
⅓   cup shredded low-fat Monterey Jack cheese
¼   cup canned diced green chili peppers
2   scallions, minced
1½ cups reduced-sodium tomato puree
1   teaspoon chili powder
¼   teaspoon ground cumin
10 frozen corn tortillas (6" diameter), thawed
¼   cup shredded nonfat mozzarella cheese
¼   cup nonfat sour cream
¼   cup chopped fresh cilantro

✸ Preheat the oven to 425°F. Coat a 13" × 9" baking dish with no-stick spray.

✸ In a medium bowl, combine the beans, tofu, and salsa. Mash together with a fork. Add the Monterey Jack, chili peppers, and scallions.

✸ In a small bowl, combine the tomato puree, chili powder, and cumin. Mix well.

✸ Spoon half of the bean mixture into the prepared baking dish. Top with 5 of the tortillas. Spoon half of the tomato mixture over the tortillas. Repeat to use the remaining bean mixture, tortillas, and tomato mixture.

✸ Top with the mozzarella, sour cream, and cilantro. Cover and bake for 15 minutes. Uncover and bake for 15 minutes, or until bubbly. Let stand for 5 minutes before serving.

Makes 6 servings

Per serving
Calories 248
Total fat 1.6 g.
Saturated fat 0.1 g.
Cholesterol 1 mg.
Sodium 427 mg.
Fiber 8.6 g.

Cost per serving

69¢

KITCHEN TIP

To freeze, cool the cooked casserole. Wrap the entire dish in freezer-quality plastic wrap, then in freezer-quality foil. To use, thaw overnight in the refrigerator. Remove the foil. Microwave on high power for 5 minutes, or until hot.

171

Pasta, Vegetable, and Egg Main Dishes

Per serving
Calories 217
Total fat 5.7 g.
Saturated fat 2.7 g.
Cholesterol 15 mg.
Sodium 439 mg.
Fiber 8 g.

Cost per serving

83¢

EGGPLANT ROLLS

If desired, serve Marinara Sauce (page 252) with these easy eggplant rolls—a modern twist on traditional pasta dishes.

2 large eggplants
1 cup frozen chopped onions
1 cup frozen chopped sweet red peppers
1 teaspoon olive oil
1 cup frozen defatted Chicken Stock
    (page 61), thawed
½ cup minced fresh parsley
½ cup shredded nonfat mozzarella cheese
½ cup nonfat ricotta cheese
½ cup grated Parmesan cheese

Preheat the broiler. Coat 2 large baking sheets with no-stick spray.

With a serrated knife, cut the eggplants lengthwise into ¼"-thick slices. Place on the prepared baking sheets. Coat with no-stick spray. Broil 4" from the heat for 5 minutes, or until golden brown. Turn and cook for 3 minutes, or until golden brown. Let cool.

In a 10" no-stick skillet, combine the onions, peppers, and oil. Cook, stirring, over medium-high heat for 3 minutes, or until the onions are golden brown. Add ¼ cup of the stock. Cook for 5 minutes, or until the vegetables are very soft. Place in a medium bowl. Add the parsley, mozzarella, ricotta, and Parmesan. Mix well.

Preheat the oven to 350°F. Coat a 13" × 9" baking dish with no-stick spray.

Spread the cheese mixture on the eggplant slices. Roll tightly to enclose the filling. Arrange, seam side down, in a single layer in the prepared baking dish. Add the remaining ¾ cup stock. Bake for 40 minutes, or until hot.

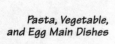

ROOT-VEGETABLE RAGOÛT

Slow oven cooking makes vegetables supersweet. This wonderful ragoût can be frozen for an easy meal on a busy day. Serve with hot bread, steamed rice, and fresh salad.

- 1 large onion, thinly sliced
- 2 carrots, diagonally sliced
- 1 tablespoon olive oil
- 2 parsnips, diagonally sliced
- 1 small rutabaga, cut into small cubes
- ¼ cup apple juice
- 4 cups peeled and cubed butternut squash
- 1 tablespoon raisins
- ¼ teaspoon ground cinnamon
- 2 cups frozen defatted Chicken Stock (page 61), thawed
- ¼ cup chopped fresh parsley
- ½ teaspoon salt
- ½ teaspoon ground black pepper

⁂ Preheat the oven to 350°F.

⁂ In a Dutch oven, combine the onions, carrots, and oil. Cook, stirring, over medium-high heat for 5 minutes, or until the onions are soft but not browned. Add the parsnips, rutabagas, and apple juice. Cook, stirring, for 3 minutes.

⁂ Add the squash, raisins, and cinnamon. Cook, stirring, for 1 minute. Add the stock. Bring to a boil.

⁂ Cover and transfer to the oven. Bake for 25 to 30 minutes, or until the squash is very soft. Add the parsley, salt, and pepper. Mix well.

Makes 4 servings

Per serving
Calories 199
Total fat 4.2 g.
Saturated fat 0.7 g.
Cholesterol 3 mg.
Sodium 314 mg.
Fiber 9.2 g.

Cost per serving

64¢

KITCHEN TIPS

To freeze, pack the cooled cooked vegetables in a freezer-quality plastic container. To use, thaw overnight in the refrigerator. Microwave on high power for 5 minutes, or until hot.

To make it easier to peel hard-shell squash, pierce the squash several times with a sharp knife, then microwave on high power for 5 minutes, turning occasionally.

173

KITCHEN TIP

To freeze, pack the cooled filled pancakes in a freezer-quality plastic container. Top with the cooled sauce. To use, thaw overnight in the refrigerator. Microwave on high power for 8 minutes, or until hot.

These thin pancakes are filled with ratatouille, the classic French summer vegetable stew, then topped with a super easy tomato sauce.

Filling

- 1 large eggplant, peeled and diced
- 2 tomatoes, chopped
- 1 cup chopped onions
- 2 zucchini, sliced
- 1 sweet red pepper, diced
- 1 tablespoon minced fresh basil
- 1 teaspoon sugar

Pancakes

- 1 cup all-purpose flour
- 2 cups skim milk
- 2 eggs
- 1 teaspoon oil

Sauce

- 2 cups chopped tomatoes
- ½ cup chopped dry-pack sun-dried tomatoes
- 1 teaspoon minced garlic
- ¼ teaspoon salt

* *To make the filling:* In a Dutch oven, combine the eggplant, tomatoes, onions, zucchini, red peppers, basil, and sugar. Cook, stirring frequently, over medium-high heat for 25 minutes, or until very thick.

* *To make the pancakes:* Meanwhile, in a blender or food processor, combine the flour, milk, eggs, and oil. Process until smooth.

* Coat a 10″ no-stick skillet with no-stick spray and place over medium-high heat until hot. Pour in ¼ cup of the batter and swirl the skillet to thinly coat the bottom. Cook for 1 minute, or until the top is set and the bottom is golden brown. Transfer the pancake to a plate. Repeat to make a total of 12 pancakes.

* *To make the sauce:* In a blender or food processor, combine the tomatoes, sun-dried tomatoes, garlic, and salt. Process until smooth. Pour into a medium saucepan. Bring to a boil over medium-high heat. Remove from the heat.

* Reheat the filling if necessary. Divide the filling among the pancakes. Roll to enclose the filling. Serve topped with the sauce.

STUFFED BUTTERNUT SQUASH WITH VEGETABLES

This hearty dish is a cinch to put together with frozen ingredients you keep on hand. Butternut squash stores well in a cool, dry cupboard for several weeks, so buy plenty when it's on sale in the fall. Look for loose-pack frozen spinach in the supermarket; it's easy to measure without thawing.

Makes 6 servings

Per serving
Calories 210
Total fat 2.7 g.
Saturated fat 1.3 g.
Cholesterol 5 mg.
Sodium 231 mg.
Fiber 11.8 g.

Cost per serving

58¢

2	large butternut squash, halved lengthwise and seeded
1	teaspoon olive oil
½	cup frozen chopped onions
6	large cloves garlic, minced
1	cup frozen broccoli florets
1	cup frozen sliced carrots
1	cup packed frozen spinach, chopped
½	teaspoon dried sage
½	teaspoon dried thyme
½	cup soft bread crumbs
¼	cup raisins
2	teaspoons lemon juice
½	teaspoon ground black pepper
¼	teaspoon salt
½	cup shredded low-fat Cheddar cheese

✺ Preheat the oven to 350°F. Cover a large baking sheet with foil. Place the squash, cut side down, on the sheet. Bake for 45 minutes, or until tender.

✺ Warm the oil in a 10″ no-stick skillet over medium-high heat until hot. Add the onions and garlic. Cook, stirring, over medium-high heat for 5 minutes, or until the onions are soft. Add the broccoli, carrots, spinach, sage, and thyme. Cook, stirring frequently, for 2 to 3 minutes, or until the vegetables soften slightly. Remove from the heat. Stir in the bread crumbs, raisins, lemon juice, pepper, and salt.

✺ Pack the vegetable mixture into the squash cavities. Top with the Cheddar. Place on the prepared baking sheet. Bake for 25 minutes, or until the topping is golden brown.

Pasta, Vegetable, and Egg Main Dishes

Per serving
Calories 285
Total fat 8.7 g.
Saturated fat 1.5 g.
Cholesterol 116 mg.
Sodium 558 mg.
Fiber 5.7 g.

Cost per serving

56¢

KITCHEN TIP

To freeze the
cooled pancakes,
place on a tray and
put in the freezer
for several hours,
or until solid.
Stack, with pieces
of wax paper
between the
pancakes, in a
freezer-quality
plastic bag. Pack
the cooled cooked
puree in a freezer-
quality plastic
container. To use,
thaw the pancakes
and puree
overnight in the
refrigerator. Place
the pancakes in a
12" no-stick skillet,
cover, and cook
over medium-high
heat for 5 minutes,
or until hot. Place
the puree in a
medium saucepan.
Cook over low heat
for 5 minutes, or
until bubbling.

*Pasta, Vegetable,
and Egg Main Dishes*

SPICY BEAN PANCAKES WITH TOMATO-PEPPER PUREE

Leftover cooked beans make moist, hearty pancakes that taste great with a Southwestern seasoned tomato sauce.

Pancakes

½ cup cooked navy or pinto beans
   (page 177)
2 eggs, lightly beaten
⅓ cup all-purpose flour
⅓ cup yellow cornmeal
1 teaspoon baking powder
½ cup low-fat buttermilk
½ cup shredded low-fat extra-sharp
   Cheddar cheese
½ small jalapeño pepper, seeded and minced
   (wear plastic gloves when handling)
1 teaspoon olive oil

Puree

1½ cups reduced-sodium tomato puree
1 sweet red pepper, minced
2 tablespoons lime juice
1 tablespoon minced garlic
1 tablespoon chopped fresh cilantro or parsley
1 teaspoon honey
½ teaspoon dried thyme
½ teaspoon salt
½ teaspoon ground black pepper

◉ *To make the pancakes:* Place the beans in a medium bowl and mash well with a fork. Add the eggs and mix well. Add the flour, cornmeal, and baking powder. Mix well. Add the buttermilk, Cheddar, jalapeño peppers, and oil. Stir until the mixture resembles a thick sauce.

◉ Coat a 10″ no-stick skillet with no-stick spray and place over medium-high heat until hot. Add 2 tablespoons of the batter. Cook for 1 to 2 minutes, or until bubbles appear on the surface of the pancake. Turn and cook for 2 minutes, or until golden brown. Place on a plate; cover to keep warm. Repeat with the remaining batter to make a total of 12 pancakes.

❁ *To make the puree:* In a blender or food processor, combine the tomato puree, red peppers, lime juice, and garlic. Process until smooth. Transfer to a medium saucepan.

❁ Stir in the cilantro or parsley, honey, thyme, salt, and black pepper. Cook, stirring occasionally, over medium heat for 10 minutes. Serve with the pancakes.

BEANS ARE BEST FOR FREEZING

Each cup of beans that you cook and freeze saves you up to 15¢ over canned. Beans are easy to cook in quantity, cool, and package into freezer containers. Freeze them for up to six months with no loss of flavor.

Use the chart below for cooking times and yields of beans called for in the recipes in this chapter.

Type (1 cup dried)	Water (cup)	Cooking Time (hr)	Yield (cup)
Black beans	4	1½	2
Chickpeas	4	2–3	2
Kidney beans	3	1½	2
Lentils	3	1	2
Lima beans	2	1½	1½
Navy beans	3	1½	2
Pinto beans	3	2½	2
Split peas	3	1	2

SWEET POTATOES STUFFED WITH CHILI

Black-bean chili looks stunning spooned over bright orange sweet potatoes. This simple entrée is rich in fiber and vitamins and costs only 83¢ a serving. If you have fresh cilantro on hand, mince some for a pretty garnish.

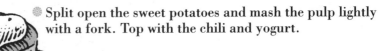

- 4 large sweet potatoes
- ¼ cup dry sherry or apple juice
- 1 teaspoon olive oil
- ½ cup frozen chopped onions
- 2 teaspoons minced garlic
- ½ cup frozen sliced carrots
- ¼ cup salsa
- 3 cups frozen cooked black beans (page 177), thawed
- 1 can (14 ounces) reduced-sodium whole tomatoes, chopped (with juice)
- ½ cup frozen peas
- 1 tablespoon chili powder
- 1 teaspoon ground cumin
- ¼ teaspoon salt
- ½ cup nonfat plain yogurt

✤ Preheat the oven to 400°F.

✤ Pierce the sweet potatoes several times with a fork. Place on a large baking sheet. Bake for 1 hour, or until very tender.

✤ In a 10″ no-stick skillet, combine the sherry or apple juice and oil. Bring to a boil over medium-high heat. Add the onions and garlic. Cook, stirring, for 3 minutes. Add the carrots and salsa. Cook, stirring, for 5 to 7 minutes, or until the vegetables soften.

✤ Add the beans, tomatoes (with juice), peas, chili powder, cumin, and salt. Reduce the heat to low. Cover and cook, stirring frequently, for 20 minutes, or until the chili is thick.

✤ Split open the sweet potatoes and mash the pulp lightly with a fork. Top with the chili and yogurt.

Freezing Dairy Products and Eggs

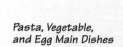

You may be surprised to learn that many dairy products can be frozen, so take advantage of good sales and follow these guidelines.

- *Butter, reduced-calorie.* Freeze in the original carton for up to 4 months.

- *Cheese, grated.* Freeze Parmesan and Romano in freezer-quality plastic containers or plastic bags for up to 2 years.

- *Cheese, ricotta.* Freeze in the original carton for up to 1 month. Ricotta can separate after thawing, so it's best used in cooked dishes. Puree in a blender or food processor, if needed.

- *Cheese, shredded.* Freeze Cheddar, Monterey Jack, mozzarella, and Swiss in small portions (1 to 2 cups) in plastic containers for up to 4 months.

- *Egg whites.* Freeze in ice-cube trays (one egg white per hole) for 9 to 12 months. Cook the yolks for pet food, if desired.

- *Sour cream, low-fat.* Freeze in the original carton for up to 4 months for use in cooking. Sour cream tends to separate after thawing. Stir well before using.

- *Yogurt.* Freeze in the original carton for up to 1 month for use in cooking. Yogurt tends to separate after thawing. Stir well before using.

Per serving
Calories 194
Total fat 7.4 g.
Saturated fat 2.8 g.
Cholesterol 222 mg.
Sodium 507 mg.
Fiber 4.7 g.

Cost per serving

88¢

VEGETABLE-CHEESE ROLL

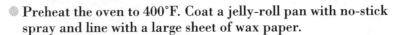

This rolled omelet is so simple to prepare, yet it makes a stunning presentation at a weekend brunch. The secret is lining the baking pan with wax paper, which makes it easier to roll the omelet after baking.

4   cups packed frozen loose-pack spinach
4   eggs, separated
1   cup shredded nonfat mozzarella cheese
¼ cup grated Parmesan cheese
¼ cup nonfat plain yogurt

FROM the Freezer

❋ Preheat the oven to 400°F. Coat a jelly-roll pan with no-stick spray and line with a large sheet of wax paper.

❋ Place the spinach in a colander. Rinse under hot running water for 30 seconds. Squeeze any excess water from the spinach. Place in a medium bowl. Add the egg yolks and mix well.

❋ Place the egg whites in a large bowl. Beat with an electric mixer until stiff peaks form. Fold the egg whites into the spinach mixture. Spread evenly in the prepared jelly-roll pan. Cover with foil. Bake for 20 minutes, or until firm.

❋ In a small bowl, combine the mozzarella, Parmesan, and yogurt. Mix well.

❋ Turn the omelet onto a clean sheet of wax paper, then peel off the lining. Spread evenly with the cheese mixture. Roll up the omelet. Cut into 8 slices.

VEGETABLE PATTIES

Hearty mushrooms boost the flavor of these tasty vegetable burgers. Allow 3 to 4 patties per serving. This recipe makes a nice large batch to freeze for quick future meals.

3   cups chopped mushrooms
¼   cup frozen defatted Chicken Stock (page 61), thawed
¼   cup chopped onions
2   teaspoons minced garlic
2   cups shredded potatoes
2   cups chopped broccoli florets
1   cup shredded low-fat Swiss cheese
¼   cup grated Parmesan cheese
2   eggs, lightly beaten
1   teaspoon dried thyme
½   teaspoon salt
¾   cup dry bread crumbs

* In a 10″ no-stick skillet, combine the mushrooms, stock, onions, and garlic. Cook, stirring, over medium-high heat for 10 minutes, or until the mushrooms release liquid.

* Add the potatoes and broccoli. Cook, stirring, for 5 minutes, or until the broccoli is soft. Transfer to a medium bowl.

* Add the Swiss, Parmesan, eggs, thyme, and salt to the bowl. Mix well. Form the mixture into 24 small patties.

* Place the bread crumbs on a plate. Dip the patties in the bread crumbs, patting to make sure the crumbs adhere.

* Wash and dry the skillet. Coat with no-stick spray and place over medium-high heat until hot. Add the patties, working in batches. Cook for 5 minutes. Turn and cook for 5 minutes, or until golden brown.

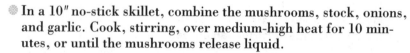

Makes 24

Per patty
Calories 57
Total fat 1.6 g.
Saturated fat 0.7 g.
Cholesterol 20 mg.
Sodium 126 mg.
Fiber 1 g.

Cost per serving

14¢

KITCHEN TIP

To freeze, place the cooled cooked patties on a tray. Set in the freezer for several hours, or until solid. Pack in a freezer-quality plastic bag. To use, thaw overnight in the refrigerator. Place in a 12″ no-stick skillet, cover, and cook over medium-high heat for 5 minutes, or until hot.

Pasta, Vegetable, and Egg Main Dishes

Per serving
Calories 256
Total fat 9.2 g.
Saturated fat 4.7 g.
Cholesterol 124 mg.
Sodium 408 mg.
Fiber 7.9 g.

Cost per serving

60¢

Garden Frittata

With a variety of frozen vegetables tucked in the freezer, you can pull together an almost effortless meatless main dish. Select loose-pack vegetables, such as the artichoke hearts, whenever possible. There's no need for thawing to measure them. If you like, you can add ½ teaspoon dried thyme, oregano, or basil to the vegetables for additional flavor.

1	teaspoon olive oil
2	cups frozen chopped onions
3	cloves garlic, minced
2	cups frozen broccoli florets
1	package (10 ounces) frozen artichoke hearts, chopped
1	cup frozen sliced carrots, chopped
1	cup shredded low-fat extra-sharp Cheddar cheese
½	cup skim milk
4	egg whites, lightly beaten
2	eggs, lightly beaten
⅛	teaspoon crushed red-pepper flakes
2	tablespoons grated Parmesan cheese

FROM the Freezer

❉ Preheat the oven to 325°F.

❉ Coat a 10" no-stick ovenproof skillet with no-stick spray and place over medium-high heat until hot. Add the oil, onions, and garlic. Cook, stirring, for 5 minutes, or until the onions are golden brown. Add the broccoli, artichoke hearts, and carrots. Cook, stirring, for 3 minutes, or until the broccoli is bright green.

❉ In a medium bowl, combine the Cheddar, milk, egg whites, eggs, and red-pepper flakes. Mix well. Add to the skillet and stir once. Cook for 3 minutes, or until the eggs begin to set on the bottom.

❉ Cover and bake for 15 to 20 minutes, or until the frittata is firm and golden brown. Sprinkle with the Parmesan.

CHILI-CHEESE QUICHE

The way to avoid a lot of fat in a quiche is to eliminate the crust. Going crustless also saves you 8¢ a serving. This lively Southwestern treatment of the French classic is a wonderful brunch or light supper dish.

1 cup chopped onions
2 teaspoons olive oil
2 cloves garlic, minced
1 cup sliced carrots
1 cup sliced mushrooms
1 cup diced tomatoes
½ cup diced scallions
½ teaspoon ground cumin
¼ teaspoon ground coriander
1 cup shredded low-fat extra-sharp Cheddar cheese
4 eggs
½ cup nonfat sour cream
1 tablespoon all-purpose flour
1 can (12 ounces) whole mild green chili peppers
¼ cup grated Parmesan cheese

❋ Preheat the oven to 375°F. Coat a 9″ freezer-proof pie pan with no-stick spray.

❋ In a 10″ no-stick skillet, combine the onions, oil, and garlic. Cook, stirring, over medium-high heat for 2 minutes. Add the carrots, mushrooms, tomatoes, scallions, cumin, and coriander. Cook, stirring, for 8 to 10 minutes, or until the vegetables are very soft.

❋ In a blender or food processor, combine the Cheddar, eggs, sour cream, and flour. Process until smooth.

❋ Slit the chili peppers lengthwise (wear plastic gloves when handling) and arrange on the bottom and up the sides of the prepared pie plate, overlapping slightly. Spoon the vegetable mixture over the peppers. Pour the egg mixture over the vegetables. Sprinkle with the Parmesan.

❋ Bake for 30 minutes, or until the quiche is golden brown and firm.

Makes 6 servings

Per serving
Calories 195
Total fat 9 g.
Saturated fat 4.1 g.
Cholesterol 155 mg.
Sodium 477 mg.
Fiber 3.9 g.

Cost per serving

98¢

KITCHEN TIP

To freeze, cool the quiche. Wrap the entire pie pan with freezer-quality plastic wrap, then with freezer-quality foil. To use, thaw overnight in the refrigerator. Remove the foil and plastic wrap. Bake at 375°F for 20 minutes, or until hot.

Pasta, Vegetable, and Egg Main Dishes

VEGETABLE SIDE DISHES

Every summer, fresh vegetables flood markets, farm stands, and backyard gardens, creating succulent selections for dinner menus. When everything is ripe at once, freezing bargains gives you the joy of garden vegetables year-round. Roasting and freezing surplus sweet peppers, for example, can save you 75 cents or more per pepper in off-season recipes.

Many chefs vote for frozen vegetables over canned, since freezing retains brighter color and crisper texture. The key to preserving the flavor and color in vegetables that are freezer-bound is to roast or blanch them first. The brief heat of an oven or boiling water inactivates the enzymes that destroy flavor and lead to loss of quality.

Many nutrition experts also applaud the nutrient content of vegetables that are frozen right after harvest. These vegetables often score higher in vitamin and mineral content than fresh vegetables that can be weeks old by the time they are eaten.

SESAME-CURRY CAULIFLOWER

Curry powder and dark sesame oil create extraordinary flavor for this winter-vegetable side dish.

4 cups frozen cauliflower florets
⅓ cup frozen defatted Chicken Stock
     (page 61), thawed
3 cloves garlic, minced
1 teaspoon dark sesame oil
1 teaspoon curry powder
½ teaspoon sugar
2 tablespoons minced fresh cilantro or parsley
⅛ teaspoon crushed red-pepper flakes
⅛ teaspoon salt
⅛ teaspoon ground black pepper

❋ In a 10″ no-stick skillet over medium-high heat, combine the cauliflower, stock, garlic, oil, curry powder, and sugar. Bring to a boil. Cover and cook, stirring frequently, for 5 minutes, or until the cauliflower is crisp-tender.

❋ Add the cilantro or parsley, red-pepper flakes, salt, and black pepper. Toss to combine.

BROCCOLI AND CARROT SAUTÉ

This bright side dish comes straight from the freezer at only 10¢ a serving— even less if you've frozen your own vegetables during harvest season.

2 cups frozen broccoli florets
2 cups frozen sliced carrots
1 cup frozen chopped onions
2 tablespoons minced garlic
2 tablespoons lemon juice
2 tablespoons chopped fresh parsley
2 teaspoons olive oil
¼ teaspoon salt
¼ teaspoon ground black pepper

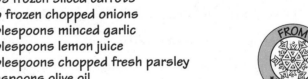

❋ Coat a 10″ no-stick skillet with no-stick spray and place over medium-high heat until hot. Add the broccoli, carrots, onions, and garlic. Cook, stirring, for 8 to 10 minutes, or until the vegetables are crisp-tender. Add the lemon juice, parsley, oil, salt, and pepper. Toss well.

GREEN BEANS WITH MUSHROOMS AND WALNUTS

Quick and pretty, this side dish adds color and crunch to a roasted chicken or pork entrée. Toasting the walnuts heightens their earthy quality.

2 tablespoons frozen defatted Chicken Stock (page 61), thawed
2 teaspoons olive oil
2 cups sliced mushrooms
1 teaspoon minced garlic
1 pound frozen green beans
2 teaspoons chopped walnuts, toasted
1 tablespoon chopped fresh parsley
¼ teaspoon salt
¼ teaspoon ground black pepper

✳ In a 10″ no-stick skillet, combine the stock and oil. Bring to a boil over medium-high heat. Add the mushrooms and garlic. Cook, stirring, for 5 to 8 minutes, or until the mushrooms are golden brown. Add the beans and walnuts. Cook, stirring, for 3 to 5 minutes, or until the beans are crisp-tender. Stir in the parsley, salt, and pepper.

MAKE THE MOST OF MUSHROOMS

Raw mushrooms don't freeze well because they absorb so much moisture from the cold freezer air. Cooked mushrooms, however, do very well in the freezer. When you get a good buy on mushrooms, keep them for months by cooking and freezing.

Coat a 10″ no-stick skillet with no-stick spray and add chopped or sliced mushrooms. Cook over medium-high heat, stirring frequently, for 10 minutes, or until the mushrooms begin to exude moisture. This is the point where their flavor intensifies. Let cool before packing into small freezer-quality containers or freezer-quality plastic bags. Cooked mushrooms can be frozen for 6 months and make delightful additions to soups, stews, sauces, and entrées.

FROM the Freezer

Makes 4 servings

Per serving
Calories 64
Total fat 3.3 g.
Saturated fat 0.4 g.
Cholesterol 0 mg.
Sodium 148 mg.
Fiber 4.1 g.

Cost per serving

54¢

KITCHEN TIP

When walnuts are on sale, stock your freezer. Pack in small freezer-quality plastic bags and store for up to 1 year. Because frozen walnuts don't clump together, it's easy to remove just the right amount for a recipe. There's no need to thaw before using.

187

Per serving
Calories 147
Total fat 7.2 g.
Saturated fat 1 g.
Cholesterol 0 mg.
Sodium 252 mg.
Fiber 2 g.

Cost per serving

57¢

Kitchen Tip

Get full value from any bunch of herbs you purchase. After using the amount you need for a recipe, chop and freeze the remaining leaves in a small freezer-quality plastic bag. Parsley and other herbs will remain green and flavorful. Use the herbs while still frozen.

Beets in Balsamic Vinegar Sauce

The red hues of this salad contrast nicely with a bed of chopped greens, such as curly endive or spinach. Most commercially frozen beets are peeled, which saves you time and red-stained hands.

2 tablespoons olive oil
2 tablespoons balsamic vinegar
1 tablespoon honey or sugar
3 cloves garlic, minced
1 teaspoon Dijon mustard
¼ teaspoon salt
¼ teaspoon ground black pepper
1 pound frozen sliced beets, thawed
1 cup frozen chopped sweet red peppers, thawed
⅓ cup chopped fresh parsley
¼ cup frozen chopped onions, thawed

* In a large bowl, combine the oil, vinegar, honey or sugar, garlic, mustard, salt, and black pepper. Mix well. Add the beets, red peppers, parsley, and onions. Mix well. Cover and refrigerate for 1 hour, stirring frequently.

CABBAGE AND PEPPER STIR-FRY

Late summer is the perfect time for this tasty stir-fry, when peppers can be had for less than 25¢ apiece. It makes a pretty accompaniment to cooked chicken or turkey breasts. If you have red cabbage on hand, substitute it for half of the green cabbage.

2   teaspoons cornstarch
¼   cup frozen defatted Chicken Stock
       (page 61), thawed
2   tablespoons packed brown sugar
2   tablespoons cider vinegar
¼   teaspoon crushed red-pepper flakes
2   teaspoons olive oil
2   large sweet red or yellow peppers, thinly sliced
3   cups sliced green cabbage
¼   cup sliced scallions

✱ Place the cornstarch in a medium bowl. Add the stock and stir until smooth. Stir in the brown sugar, vinegar, and red-pepper flakes.

✱ Coat a 10″ no-stick skillet with no-stick spray and place over medium-high heat until hot. Add the oil, peppers, and cabbage. Cook, stirring, for 3 to 5 minutes, or until crisp-tender. Add the scallions and the stock mixture. Cook, stirring, for 1 minute, or until the sauce thickens.

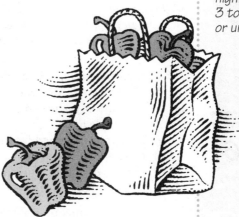

Makes 4 servings

Per serving
Calories 91
Total fat 2.6 g.
Saturated fat 0.4 g.
Cholesterol 0 mg.
Sodium 22 mg.
Fiber 3.6 g.

Cost per serving

49¢

KITCHEN TIP

To freeze, pack the cooled cooked stir-fry in a freezer-quality plastic container. To use, thaw overnight in the refrigerator. Place in a 10″ no-stick skillet and cook over medium-high heat for 3 to 5 minutes, or until hot.

CARROT AND PARSNIP PUREE

Parsnips—root vegetables that look like white carrots—are too frequently overlooked in the winter produce market. Try them in this fast puree, and you'll discover just how sweet they can be.

3 cups frozen sliced carrots
3 cups frozen sliced parsnips
2 large cloves garlic
2 tablespoons defatted Chicken Stock
   (page 61), thawed
1 teaspoon olive oil
¼ teaspoon salt

❋ Place the carrots, parsnips, and garlic in a medium saucepan and cover with water. Bring to a boil over medium-high heat. Cook for 15 minutes, or until tender but not mushy. Check by inserting the tip of a sharp knife into 1 piece of parsnip. Drain.

❋ Transfer to a blender or food processor. Add the stock, oil, and salt. Process until smooth. Transfer to the saucepan and cook, stirring constantly, over low heat for 3 minutes, or until hot.

HONEY CARROTS WITH MUSTARD

Because of their economy and versatility, carrots are a must to keep on hand in the freezer. This sweet-sharp side dish pairs well with pork. It takes less than 10 minutes to make and costs less than 20¢ a serving.

4 cups frozen sliced carrots
½ cup frozen defatted Chicken Stock
   (page 61), thawed
½ cup apple juice
½ teaspoon chopped fresh ginger
1 tablespoon honey
1 tablespoon Dijon mustard
1 clove garlic, minced

❋ In a large saucepan, combine the carrots, stock, apple juice, and ginger. Bring to a boil over medium-high heat. Cook, stirring occasionally, for 10 minutes, or until the liquid has evaporated and the carrots are tender. Check by inserting the tip of a sharp knife into 1 carrot.

❋ Stir in the honey, mustard, and garlic.

FENNEL BAKE WITH PARMESAN

Fennel has a sweet, anise flavor. Use firm, compact bulbs with crisp stalks. Cooked fennel freezes for up to 1 month.

½ cup frozen defatted Chicken Stock
    (page 61), thawed
2 bulbs fennel (1 pound each), trimmed
    and thinly sliced
2 tablespoons chopped frozen onions, thawed
¼ teaspoon dry mustard
¼ teaspoon ground black pepper
⅛ teaspoon salt
¼ cup grated Parmesan cheese

※ Bring the stock to a boil in a Dutch oven over medium-high heat. Add the fennel. Cook for 10 minutes, or until the liquid has evaporated and the fennel is tender. Check by inserting the tip of a sharp knife into 1 slice.

※ Preheat the oven to 400°F.

※ In a small bowl, combine the onions, mustard, pepper, and salt. Sprinkle over the fennel. Top with the Parmesan. Cover and bake for 25 minutes, or until golden brown.

Makes 4 servings

Per serving
Calories 104
Total fat 2.4 g.
Saturated fat 1.2 g.
Cholesterol 5 mg.
Sodium 302 mg.
Fiber 9.5 g.

Cost per serving

57¢

ASIAN GREEN BEANS

Dry-frying green beans is an Asian technique that requires scant oil—just enough for flavoring. Frozen sliced green beans cost about 69¢ a pound and work well in this tasty side dish.

1 pound frozen sliced green beans
1 teaspoon reduced-sodium soy sauce
1 teaspoon sugar
1 teaspoon dark sesame oil
1 teaspoon olive oil
2 cloves garlic, minced

※ Rinse the beans and pat them dry with paper towels. Place in a large bowl. Add the soy sauce, sugar, and sesame oil. Toss well.

※ Coat a 10″ no-stick skillet with no-stick spray and place over medium-high heat until hot. Add the olive oil and garlic. Cook, stirring, for 1 minute. Add the bean mixture. Cook, stirring, for 3 to 5 minutes, or until the beans begin to blister.

Makes 4 servings

Per serving
Calories 54
Total fat 2.4 g.
Saturated fat 0.4 g.
Cholesterol 0 mg.
Sodium 64 mg.
Fiber 3.4 g.

Cost per serving

23¢

Per serving
Calories 184
Total fat 3.7 g.
Saturated fat 0.5 g.
Cholesterol 0 mg.
Sodium 155 mg.
Fiber 2 g.

Cost per serving

41¢

KITCHEN TIP

To freeze, pack the cooled cooked mixture in freezer-quality plastic containers. To use, thaw overnight in the refrigerator. Place in a 10" no-stick skillet, cover, and cook over medium heat for 5 to 7 minutes, or until bubbling.

SWEET-AND-SOUR ONIONS AND PINEAPPLE

Onions get an intriguing Asian-style treatment in this side dish that freezes well for up to 3 months. Serve with pork chops, roasted chicken, or catfish.

1 tablespoon olive oil
4 medium onions, sliced
2 cloves garlic, minced
1 can (20 ounces) unsweetened
   pineapple chunks (with juice)
¼ cup cider vinegar
1 tablespoon reduced-sodium soy sauce
2 tablespoons cornstarch
3 tablespoons apple juice

❋ Coat a 10" no-stick skillet with no-stick spray and place over medium heat until hot. Add the oil, onions, and garlic. Cook, stirring frequently, for 15 minutes, or until the onions are very soft but not browned. Add the pineapple (with juice), vinegar, and soy sauce. Bring to a boil.

❋ Place the cornstarch in a cup. Add the apple juice and stir until smooth. Add to the skillet. Cook, stirring, for 5 minutes, or until the sauce thickens.

GARLIC GUSTO

Convenience counts in everyone's kitchen, but with chopped jarred garlic, convenience costs you 85¢ or more for every ½ cup you buy. Bank those savings and still make your recipe preparation a breeze. Buy garlic in bulk during the summer sales, then chop and freeze it.

First break the garlic bulbs into cloves. Using the flat side of a chef's knife or any other knife with a wide blade, smack each clove sharply to partially crush it and easily release the skin. Lift the cloves (the skin should fall away from the cloves) and chop finely in a food processor or food chopper.

Measure tablespoonfuls onto a baking sheet or plate lined with wax paper, then freeze. When frozen, peel the dabs from the paper and pack into freezer-quality plastic bags. Frozen garlic retains its flavor for up to 3 months.

WARM VEGETABLES
WITH CHILI-PEPPER VINAIGRETTE

Makes 4 servings

Per serving
Calories 122
Total fat 0.4 g.
Saturated fat 0.1 g.
Cholesterol 0 mg.
Sodium 200 mg.
Fiber 5.7 g.

Cost per serving

41¢

Frozen vegetables come to life in this sweet-and-spicy side dish. Because the frozen ingredients are already prepped, it takes only minutes to put together. Vary the vegetables according to what's in your freezer.

2 cups frozen sliced carrots
1 cup frozen broccoli florets
1 cup frozen chopped sweet red peppers
1 cup chopped red cabbage
½ cup frozen defatted Chicken Stock
      (page 61), thawed
½ cup chopped fresh parsley
¼ cup frozen chopped onions
3 tablespoons balsamic vinegar
3 tablespoons honey
½ small jalapeño pepper, minced
      (wear plastic gloves when handling)
1 tablespoon minced garlic
¼ teaspoon salt
¼ teaspoon ground black pepper

❋ In a 10″ no-stick skillet, combine the carrots, broccoli, red peppers, cabbage, and stock. Bring to a boil over medium-high heat. Cook, stirring occasionally, for 6 to 8 minutes, or until the broccoli is bright green. Drain in a colander.

❋ In a large bowl, combine the parsley, onions, vinegar, honey, jalapeño peppers, garlic, salt, and black pepper. Mix well. Add the vegetables and toss to coat. Let stand at room temperature for 10 minutes, stirring occasionally.

Per serving
Calories 97
Total fat 1.6 g.
Saturated fat 0.2 g.
Cholesterol 0 mg.
Sodium 148 mg.
Fiber 4.1 g.

Cost per serving

23¢

SUCCOTASH

This recipe combines old-fashioned flavor with up-to-the-minute speed. Garlic and paprika bring this dish a surprising boost of flavor.

1 teaspoon olive oil
1 cup frozen lima beans
1 cup frozen whole kernel corn
1 tablespoon frozen chopped onions
2 cloves garlic, minced
2 tablespoons chopped fresh chives or scallions
¼ teaspoon salt
⅛ teaspoon paprika

❋ Coat a 10″ no-stick skillet with no-stick spray and place over medium-high heat until hot. Add the oil, lima beans, corn, onions, and garlic. Cook, stirring, for 8 to 10 minutes, or until the lima beans are tender. Stir in the chives or scallions, salt, and paprika.

GROW AND FREEZE FRESH HERBS

Save up to $2 a bunch when you grow your own fresh herbs and freeze them for fall or winter use. Frozen chopped herbs taste and look nearly like fresh. Herbs grow great in pots on the patio, in window boxes, and in the garden. They like plenty of sun and an equal mixture of sand, potting soil, and peat moss. Some experts say that the poorer the soil, the better the herbs' flavor.

Harvest fresh herbs in the early morning, after the dew has dried but before the sun is hot. If they need it, wash them and then dry between paper towels. Chop finely. Portion them in small freezer-quality plastic bags. Be sure to label and date the bags. Freeze for up to four months.

Great herbs to have on hand in the freezer include basil, cilantro, oregano, parsley, rosemary, sage, savory, and thyme. Basil, cilantro, and parsley are especially good choices because the dried versions lack the vibrant taste and color of the fresh leaves.

CORN CUSTARD

Cold-weather comfort foods like this creamy custard are staples of midwestern harvest dinners. Baking the custard in a hot-water bath keeps it from browning too much. This recipe saves you 13¢ a serving over commercial versions.

1 cup evaporated skim milk
3 eggs
1 egg white
2 tablespoons frozen chopped onions
½ teaspoon salt
½ teaspoon ground black pepper
4 cups frozen whole kernel corn

❋ Preheat the oven to 325°F. Coat an 8″ × 8″ baking dish with nostick spray.

❋ In a blender or food processor, combine the milk, eggs, egg white, onions, salt, pepper, and 2 cups of the corn. Process until smooth.

❋ Transfer to a large bowl. Stir in the remaining 2 cups corn. Pour into the prepared baking dish.

❋ Place the baking dish in a larger ovenproof pan. Add hot water to the larger pan to a depth of 1″. Bake for 1¼ hours, or until a knife inserted in the center of the custard comes out clean.

Makes 4 servings

Per serving
Calories 245
Total fat 4 g.
Saturated fat 1.3 g.
Cholesterol 162 mg.
Sodium 409 mg.
Fiber 4.1 g.

Cost per serving

58¢

FREEZER-FRESH VEGETABLES

Who says the early bird always catches the worm? You can get great bargains on vegetables at the farmers' market if you shop near the end of the day, when sellers are eager to pack up and go home. Sometimes a bushel of green beans, broccoli, or carrots can be yours for less than $5. If your freezer is small, split the bounty with friends.

At home, wash and trim the vegetables. Most vegetables need to be blanched, an easy technique that inactivates vegetables' enzymes so that they don't lose quality in the freezer.

To blanch, bring a large pot of water to a boil, then immerse the vegetables for the required time. Drain and rinse under cold water. Drain again and pat dry. Pack into desired servings in freezer-quality containers or plastic bags.

Keep optimum flavor, color, and nutrients in your frozen vegetables by removing as much air as possible from the package before freezing. When too much air remains in the freezer package, it can cause freezer burn—an undesirable drying and discoloration of the food's surface. The air holds moisture, which freezes into ice crystals that leach vegetables' vitality. Press out as much excess air as possible before sealing the bag. Fill a rigid container nearly to the top. Label and date, then freeze for up to six months.

Smart summer shopping and an hour in your kitchen yield plenty of good eating from the freezer throughout the winter months.

Asparagus. Leave whole or cut into thirds. Blanch in boiling water for 30 seconds, cool, then pack into freezer-quality plastic bags.

Beans, green. Leave whole or cut into thirds. Blanch in boiling water for 3 minutes, cool, then pack into freezer-quality plastic bags.

Beets. Steam whole baby beets or sliced beets for 10 minutes, or until tender. Slip off the peels and discard. Cool, then pack into freezer containers.

Broccoli. Break into florets. Blanch in boiling water for 3 minutes, cool, then pack into freezer-quality plastic bags.

Cabbage. Slice or chop coarsely. Blanch in boiling water for 3 minutes, cool, then pack into freezer-quality plastic bags.

Carrots. Slice. Blanch in boiling water for 2 minutes, cool, then pack into freezer-quality plastic bags.

Cauliflower. Break into florets. Blanch in boiling water for 4 minutes, cool, then pack into freezer-quality plastic bags.

Corn. Blanch whole ears in boiling water for 3 minutes, cool, then cut kernels off the cobs. Pack into freezer-quality plastic bags.

Peas. Blanch in boiling water for 30 seconds, cool, then pack into freezer-quality plastic bags.

Peppers, sweet. Halve and roast (see page 267), cool, then pack into freezer-quality plastic bags.

Tomatoes. Plunge the tomatoes into boiling water for 30 seconds, then remove with a slotted spoon into a sinkful of cold water. Peel, then pack whole in freezer containers.

Per serving
Calories 257
Total fat 2.4 g.
Saturated fat 1.6 g.
Cholesterol 10 mg.
Sodium 166 mg.
Fiber 4.5 g.

Cost per serving

49¢

CREAMY POTATO SALAD WITH CORN AND PEAS

Dress up plain potato salad for pennies with an assortment of vegetables from the freezer.

5 cups cubed potatoes
¾ cup low-fat sour cream
¼ cup nonfat mayonnaise
¼ cup minced fresh parsley
2 tablespoons minced fresh basil
½ teaspoon ground black pepper
½ cup frozen whole kernel corn, thawed
½ cup frozen peas, thawed
¼ cup frozen chopped sweet red peppers, thawed

❁ Bring a large pot of water to a boil over medium-high heat. Add the potatoes. Cook for 10 minutes, or until the potatoes are tender but not mushy. Drain in a colander.

❁ In a large bowl, combine the sour cream, mayonnaise, parsley, basil, and black pepper. Mix well. Add the corn, peas, red peppers, and potatoes. Toss well to coat. Cover and refrigerate for 20 minutes, or until chilled.

POTATO APPEAL

If you peel your russet, or baking, potatoes, be sure to save the skins. They make a quick snack that's impossible to resist—and you'll save 25¢ an ounce over commercial potato chips.

Scrub the potatoes well, then peel them slightly thicker than normal, trying to get large pieces. Place the potato peels in a bowl and coat with no-stick spray. Sprinkle with ground black pepper, grated Parmesan cheese, and a pinch of salt; toss well to coat. Coat a large baking sheet with no-stick spray. Spread the peels on the baking sheet. Bake at 400°F for 10 minutes, or until crisp and golden brown. Eat right away or refrigerate in a plastic container. To crisp, place on a baking sheet in a 400°F oven for 5 to 8 minutes.

STUFFED BAKED ONIONS

*August is the most economical time to make this traditional Mediter-
ranean onion dish. You can take advantage of garden bounty to make
extra portions to serve with beef or pasta dishes.*

  4 large onions
  1 cup soft bread crumbs
  ½ cup chopped tomatoes
  ¼ cup chopped sweet red peppers
  ¼ cup chopped celery
  ¼ cup raisins
  2 tablespoons balsamic vinegar
  ½ teaspoon dried oregano
  ⅓ cup frozen defatted Chicken Stock
     (page 61), thawed
  1 tablespoon olive oil

✳ Preheat the oven to 375°F.

✳ Bring a large pot of water to a boil over high heat. Peel the
  onions and slice ½″ off the top of each. Place in the pot. Cover
  and cook for 3 minutes, or until slightly softened. Drain in a
  colander. Let stand until cool enough to handle.

✳ With a sharp spoon or melon baller, scoop the interior from
  each onion, leaving a 1″ shell. Reserve the center portions for
  another use.

✳ In a medium bowl, combine the bread crumbs, tomatoes, peppers, celery, raisins, vinegar, and oregano. Stuff
  into the onion cavities. Place the onions in an 8″
  × 8″ baking dish. Pour the stock around the
  onions. Drizzle with the oil.

✳ Cover and bake for 1 hour,
  or until the onions are soft.
  Check by inserting the tip of
  a sharp knife into 1 onion.

Makes 4 servings

Per serving
Calories 145
Total fat 4.3 g.
Saturated fat 0.7 g.
Cholesterol 1 mg.
Sodium 73 mg.
Fiber 2.7 g.

Cost per serving

39¢

KITCHEN TIP

*To freeze, pack the
cooled cooked
onions in a single
layer in a freezer-
quality plastic
container. To use,
thaw overnight in
the refrigerator.
Cover and
microwave on
high power for 3 to
5 minutes, or
until hot.*

Per serving
Calories 111
Total fat 3.9 g.
Saturated fat 0.6 g.
Cholesterol 0 mg.
Sodium 141 mg.
Fiber 4.6 g.

Cost per serving

97¢

KITCHEN TIP

Get full value from any bunch of herbs you purchase. After using the amount you need for a recipe, chop and freeze the remaining leaves in a small freezer-quality plastic bag. Basil and other herbs will remain green and flavorful. Use the herbs while still frozen.

SNOW PEAS WITH BASIL

Snow peas are a delicate vegetable often used in Asian cooking. They freeze well and are available in most supermarkets. This colorful side dish can dress up broiled fish or chicken.

2 teaspoons cider vinegar
2 teaspoons chopped fresh basil
3 teaspoons olive oil
1 cup frozen chopped sweet red peppers
¼ cup frozen whole kernel corn
¼ teaspoon salt
¼ teaspoon ground black pepper
3 cups frozen snow peas

❋ In a large bowl, combine the vinegar, basil, and 2 teaspoons of the oil.

❋ Coat a 10″ no-stick skillet with no-stick spray and place over medium-high heat until hot. Add the remaining 1 teaspoon oil. Add the red peppers and corn. Cook, stirring, for 2 minutes, or until the peppers soften slightly. Sprinkle with the salt and black pepper.

❋ Add the snow peas. Cook, stirring, for 3 minutes, or until the snow peas are crisp-tender. Transfer to the bowl. Toss well to coat with the vinegar mixture.

IMPROMPTU SALADS FROM THE FREEZER

Although you can't freeze lettuce successfully, you can toss together an almost-instant salad from other frozen foods. Carrots, peas, broccoli, cauliflower, and other vegetables can be thawed speedily by placing them in a colander in the sink and running hot water over them. If you're going to cook the vegetables, you needn't thaw them first. All of these quick dishes make four servings.

- *Lemony Broccoli Salad.* Steam 3 cups frozen broccoli florets for 5 minutes; drain well. Place in a medium bowl. Add ¼ cup minced thawed frozen sweet red peppers and 1 teaspoon toasted chopped walnuts; toss. Sprinkle with 2 tablespoons lemon juice and ¼ teaspoon salt. Mix well.

- *Marinated Onion Salad.* In a medium bowl, toss together 2 cups thawed frozen pearl onions, ¼ cup minced fresh parsley, 2 tablespoons balsamic vinegar, and ½ teaspoon sugar. Let stand for 5 minutes at room temperature, stirring occasionally.

- *Pea and Corn Salad.* In a medium bowl, toss together 2 cups thawed frozen peas, 1 cup thawed frozen corn, 2 tablespoons balsamic vinegar, 1 tablespoon olive oil, 1 tablespoon minced fresh parsley or cilantro, and ½ teaspoon minced garlic.

Per serving
Calories 34
Total fat 1.3 g.
Saturated fat 0.2 g.
Cholesterol 0 mg.
Sodium 136 mg.
Fiber 0.9 g.

Cost per serving

56¢

KITCHEN TIP

To freeze, place
the cooled cooked
peppers in a single
layer on a baking
sheet covered with
wax paper. Put in
the freezer for
1 hour, or until
solid. Peel off the
paper and
transfer to a
freezer-quality
plastic bag. To
use, thaw
overnight in the
refrigerator. Cover
and microwave on
high power
for 3 to 5
minutes, or
until hot.

OVEN-ROASTED PEPPERS

*Roasted and peeled, sweet peppers freeze for up to 3 months without losing
their vivid color, flavor, or texture. If you put the peppers in a paper bag
for about 10 minutes just after roasting, they'll be very easy to peel.*

4 sweet red or yellow peppers, halved
1 tablespoon balsamic vinegar
1 teaspoon olive oil
½ teaspoon lemon juice
¼ teaspoon salt
¼ teaspoon ground black pepper

❋ Preheat the oven to 450°F. Place the red or yellow peppers, cut
side down, on a no-stick baking sheet.

❋ In a cup, mix the vinegar and oil. Brush over the peppers. Bake
for 20 to 30 minutes, or until the skins blister. Transfer the peppers to a paper bag. Close the bag and let stand for 10 minutes.

❋ Remove the peppers from the bag. Rinse the peppers under cold
water and rub to remove the skin; discard the skin. Place the
peppers in a medium bowl. Sprinkle with the lemon juice, salt,
and black pepper. Toss to combine.

SPANISH STUFFED TOMATOES

In summer, make this easy Spanish side dish with garden-ripe tomatoes. Frozen vegetables make the assembly quick. Use a serrated knife to make quick work of cutting the frozen vegetables.

- 4 large tomatoes
- 1 teaspoon olive oil
- 5 cloves garlic, minced
- ½ cup frozen artichoke hearts, minced
- ⅓ cup frozen broccoli florets, minced
- 1 cup soft bread crumbs
- ½ cup crumbled feta cheese
- 2 teaspoons dried Italian herb seasoning
- 1 drop hot-pepper sauce
- 2 tablespoons grated Parmesan cheese

❀ Preheat the oven to 350°F.

❀ Slice off the tops of the tomatoes; discard the tops. With a grapefruit spoon or melon baller, scoop out the interior, leaving a 1″ shell. Reserve the tomato interior for another use. Place the tomatoes, cut side up, on a baking sheet.

❀ Coat a 10″ no-stick skillet with no-stick spray and place over medium-high heat until hot. Add the oil and garlic. Cook, stirring, for 1 minute. Add the artichoke hearts and broccoli. Cook, stirring, for 3 minutes, or until the broccoli is slightly softened. Remove from the heat.

❀ Add the bread crumbs, feta, Italian herb seasoning, and hot-pepper sauce. Mix well.

❀ Stuff the tomatoes with the bread-crumb mixture. Top with the Parmesan. Bake for 20 minutes, or until the topping is golden brown.

Makes 4 servings

Per serving
Calories 195
Total fat 9.9 g.
Saturated fat 5.6 g.
Cholesterol 30 mg.
Sodium 490 mg.
Fiber 3.8 g.

Cost per serving

92¢

Per serving
Calories 131
Total fat 0.4 g.
Saturated fat 0.2 g.
Cholesterol 1 mg.
Sodium 81 mg.
Fiber 2.8 g.

Cost per serving

30¢

KITCHEN TIP

To freeze, pack the
cooled cooked
puree into a
freezer-quality
plastic container.
To use, thaw
overnight in the
refrigerator. To
reheat, cover and
microwave on
high power for 3 to
5 minutes, or
until hot.

GINGERY SWEET POTATOES

*Prepare this holiday side dish ahead of time and freeze for up to
2 months. Cooking on the big day will be that much easier. The fresh
ginger lends a sweet-spicy note. If you don't have fresh ginger, you can
use 1 teaspoon dried ground ginger.*

3 large sweet potatoes
2 tablespoons frozen orange juice
      concentrate, thawed
2 tablespoons packed brown sugar
2 teaspoons minced fresh ginger
½ teaspoon reduced-calorie butter
½ teaspoon ground cinnamon
¼ teaspoon ground nutmeg
⅛ teaspoon salt

❀ Peel the sweet potatoes and cut into 1" pieces. Place in a large
saucepan and add water to cover. Bring to a boil over medium-
high heat. Cook for 15 minutes, or until tender. Check by inserting the tip of a sharp knife into 1 piece.

❀ Drain in a colander. Transfer to a blender or food processor.
Add the orange juice concentrate, brown sugar, ginger, butter,
cinnamon, nutmeg, and salt. Process until smooth. Return to
the saucepan and heat through.

PEPPERY GREENS

Makes 4 servings

Per serving
Calories 116
Total fat 4 g.
Saturated fat 0.6 g.
Cholesterol 0 mg.
Sodium 206 mg.
Fiber 6.6 g.

Cost per serving

55¢

This easy side dish features nutritious dark leafy greens, readily available loose-packed in bags in the supermarket freezer section. If you blanch and freeze them yourself when they're on sale, you can save about 10¢ a cup over commercially frozen greens.

4 cups packed frozen chopped spinach
2 cups frozen chopped collard greens
    or Swiss chard
1 clove garlic, thinly sliced
1 tablespoon olive oil
½ teaspoon crushed red-pepper flakes
½ teaspoon ground black pepper

* In a 10" no-stick skillet over medium-high heat, combine the spinach, collard greens or Swiss chard, and garlic. Cover and cook for 5 to 7 minutes, or until tender. Drain and place in a medium bowl.

* Add the oil, red-pepper flakes, and black pepper to the skillet. Place over medium-high heat until the oil is hot. Add the greens back to the skillet and toss to coat with the oil mixture.

SOUTHWESTERN BARBECUED CORN

Makes 6 servings

Per serving
Calories 132
Total fat 2.5 g.
Saturated fat 0.4 g.
Cholesterol 0 mg.
Sodium 94 mg.
Fiber 4.2 g.

Cost per serving

19¢

You can make this zesty side dish for just 19¢ a serving during corn season. If you want to cook up extra ears for the freezer, they'll store well for up to 3 months. To use, thaw overnight in the refrigerator or thaw in the microwave. Reheat on the grill or broil 4" from the heat for 5 minutes.

2 teaspoons olive oil
½ teaspoon cider vinegar
½ teaspoon ground cumin
½ teaspoon ground paprika
¼ teaspoon garlic powder
¼ teaspoon salt
1 drop hot-pepper sauce
6 frozen ears corn

* Preheat the grill. In a small bowl, combine the oil, vinegar, cumin, paprika, garlic powder, salt, and hot-pepper sauce. Mix well.

* Place the corn on a large piece of foil. Brush the oil mixture on the corn. Fold up the edges of the foil to create a sealed packet. Place on the grill. Cook for 10 minutes, or until hot.

CHAPTER 9

GRAIN, RICE, AND PASTA SIDE DISHES

If you can boil water, you can cook grains and pasta. Store cooked grains in the freezer and you have a jump-start on quick weeknight meals. Each meal that includes these types of complex carbohydrates builds the foundation of good health, according to the dietary recommendations of the U.S. Department of Agriculture Food Guide Pyramid. Your body benefits from increased fiber, vitamins, and minerals.

Thrift experts love the freezer for these basic meal components. Freezing keeps grains and whole-grain flours fresh for up to 18 months by protecting them from insects and rancidity. Uncooked whole grains can be used directly from the freezer without thawing. Let cooked rice, barley, wild rice, and other grains thaw for several hours in the refrigerator—or thaw in the microwave—before adding them to a recipe.

Create your own on-the-run dinners by packing part of your freezer with great grain and pasta dishes.

207

BARLEY AND CORN SALAD

You'll get 40 percent of the daily requirement for fiber in just one serving of this refreshing, crunchy salad.

Per serving
Calories 317
Total fat 8.4 g.
Saturated fat 0.8 g.
Cholesterol 5 mg.
Sodium 154 mg.
Fiber 10 g.

Cost per serving
60¢

3	cups frozen defatted Chicken Stock (page 61), thawed
1	cup barley
¼	teaspoon salt
2	cups frozen whole kernel corn
¾	cup frozen green beans
1	tablespoon apple juice or water
¾	cup sliced radishes
3	tablespoons lemon juice
2	tablespoons minced fresh parsley
2	tablespoons oil
1	tablespoon chopped garlic
½	teaspoon dried basil
¼	teaspoon ground black pepper

In a medium saucepan, combine the stock, barley, and salt. Bring to a boil over medium-high heat. Reduce the heat to low, cover, and cook for 40 minutes, or until the barley is tender. Let cool.

In a 10″ no-stick skillet, combine the corn, beans, and apple juice or water. Cover and bring to a boil over medium-high heat. Drain well. Place in a large bowl.

Add the radishes, lemon juice, parsley, oil, garlic, basil, pepper, and barley. Toss well.

CORN
APPLES

BARLEY WITH PARMESAN

This hearty grain pilaf is filled with fiber and nutrients. Add strips of cooked chicken for a one-dish meal.

  1 teaspoon olive oil
  1 cup frozen chopped onions
  ½ cup frozen sliced carrots, chopped
  ½ cup diced celery
  2 tablespoons minced garlic
  4 cups frozen defatted Chicken Stock
     (page 61), thawed
1¾ cups barley
  ¼ cup grated Parmesan
  2 tablespoons chopped fresh parsley
  ¼ teaspoon salt
  ¼ teaspoon ground black pepper

❋ Coat a Dutch oven with no-stick spray and place over medium-high heat until hot. Add the oil, onions, carrots, celery, and garlic. Cook, stirring, for 5 minutes, or until the onions are golden brown. Add the stock and barley. Bring to a boil.

❋ Reduce the heat to low, cover, and cook for 40 minutes, or until the barley is tender. Stir in the Parmesan, parsley, salt, and pepper.

GRAINS MAKE CENTS

Whole grains help keep your budget lean and your body healthy. The U.S. Department of Agriculture Food Guide Pyramid recommends 6 to 11 servings a day of the healthy-foundation foods—those in the bread, cereal, rice, and pasta group. In general, says thrift expert Amy Dacyczyn, author of *The Tightwad Gazette* series, those foods that you're supposed to eat the most of tend to be the cheaper items. Breads, grains, rice, and pasta cost from 15¢ to 50¢ a pound, compared with ground beef at $1 a pound and cheese at $1.60 a pound. Shift focus from a meat-based menu to a grain-based menu and see the savings mount.

MARINATED BULGUR AND BEAN SALAD

Bulgur, partially cooked cracked wheat, is a nutritious winter staple in the Slavic countries, where this salad originates. At only 34¢ a serving, it's an easy-to-assemble side dish for fish or chicken.

Per serving
Calories 178
Total fat 3 g.
Saturated fat 0 g.
Cholesterol 2 mg.
Sodium 24 mg.
Fiber 7.9 g.

Cost per serving

34¢

- 2 cups frozen defatted Chicken Stock (page 61), thawed
- 1 cup bulgur
- 1 cup frozen cooked kidney beans (page 177), thawed
- 1 cup frozen chopped sweet red peppers, thawed
- ½ cup frozen whole kernel corn
- ½ cup frozen chopped onions
- ¼ cup frozen green beans
- 2 scallions, chopped
- 3 tablespoons balsamic vinegar
- 3 cloves garlic, minced
- 1 tablespoon minced fresh cilantro
- 1 tablespoon olive oil
- ½ teaspoon Dijon mustard
- 6 leaves lettuce

FROM
the Freezer

✸ In a medium saucepan, combine the stock and bulgur. Bring to a boil over medium-high heat. Reduce the heat to low, cover, and cook for 5 minutes. Remove from the heat. Let stand for 10 minutes, or until the liquid has been absorbed. Place in a medium bowl.

✸ To the bowl, add the kidney beans, red peppers, corn, onions, green beans, scallions, vinegar, garlic, cilantro, oil, and mustard. Toss well. Arrange on the lettuce.

POLENTA WITH SWEET POTATOES

This fluffy polenta with sweet potatoes freezes for 6 weeks with no loss of flavor and texture. At only 15¢ a serving, it's economical but special enough for company.

- 2 large sweet potatoes
- 1 cup frozen defatted Chicken Stock (page 61), thawed
- 1 cup yellow cornmeal
- 3 cups boiling water
- ½ teaspoon salt
- ½ cup shredded low-fat mozzarella cheese
- 2 tablespoons grated Parmesan cheese

※ Preheat the oven to 400°F. Pierce the sweet potatoes several times with the tip of a sharp knife. Place on a large baking sheet. Bake for 45 minutes, or until very soft. Let cool, then peel. Place in a blender or food processor and process until smooth.

※ Place the stock in a medium saucepan. Whisk in the cornmeal until smooth. Whisk in the water and salt. Cook, stirring, over medium-high heat for 5 minutes, or until the polenta thickens. Reduce the heat to low, cover, and cook, stirring frequently, for 10 minutes. Add the sweet potatoes. Stir in the mozzarella and Parmesan.

KITCHEN TIP

To freeze, pack cooled cooked polenta in a freezer-quality plastic container. To use, thaw overnight in the refrigerator. Cover and microwave on high power for 3 to 5 minutes, or until hot; stop and stir once during this time.

Grain, Rice, and Pasta Side Dishes

MILLET AND SWEET-POTATO CAKES

Millet is a grain prized in Europe for its chewy texture. In this country, it's a frequent component of bird-seed mixture. But don't let the birds get all its fiber and nutty flavor. Use it in recipes like this, where millet blends with grated sweet potatoes in savory pancakes that are an inspired match to pork or beef.

1   cup frozen defatted Chicken Stock
     (page 61), thawed
⅓   cup millet
1   small sweet potato, peeled and shredded
¼   cup minced onions
2   eggs, lightly beaten
1   tablespoon all-purpose flour
½   teaspoon dried thyme
⅛   teaspoon crushed red-pepper flakes
⅛   teaspoon salt
⅛   teaspoon ground black pepper

✻ In a small saucepan, combine the stock and millet. Bring to a boil over medium-high heat. Reduce the heat to low, cover, and cook for 25 minutes, or until the millet is tender. Place in a medium bowl.

✻ Add the sweet potatoes, onions, eggs, flour, thyme, red-pepper flakes, salt, and pepper. Mix well.

✻ Coat a 10" no-stick skillet with no-stick spray and place over medium-high heat until hot. Drop spoonfuls of the batter into the skillet. Cook for 3 to 4 minutes, or until golden brown. Carefully turn and cook for 3 minutes, or until golden brown. Transfer to a plate and cover to keep warm. Repeat to use all the batter.

Makes 8 cakes

Per cake
Calories 152
Total fat 2.1 g.
Saturated fat 0.6 g.
Cholesterol 54 mg.
Sodium 58 mg.
Fiber 2.1 g.

Cost per serving

10¢

KITCHEN TIP

To freeze the cakes, place on a tray. Put in the freezer for 1 hour, or until solid. Stack, separated by small pieces of wax paper, in a freezer-quality plastic bag. To use, thaw overnight in the refrigerator. Place in a 10" no-stick skillet, cover, and cook over low heat for 5 to 8 minutes, or until hot.

MAKE YOUR OWN QUICK-COOKING GRAINS

Most supermarkets carry quick-cooking grains, which are parboiled then dried to halve the amount of stove time. You pay up to double the cost, though. Quick-cooking brown rice costs 15¢ a serving, compared with regular brown rice at 8¢.

By planning ahead, you can cook extra of any grain. Simply place the uncooked grain in a medium saucepan with water or defatted Chicken Stock (page 61); see below for amounts. Bring to a boil, then reduce the heat to low. Cover and cook for the time indicated. Let cool completely, then pack into freezer-weight containers or plastic bags and freeze for up to six months.

Add cooked grains to salads, soups, stews, and chili. Serve them with spaghetti sauce or pesto. Dress them up with stir-fried vegetables. No matter how you use them, you get extra protein and fiber in your meals.

Enjoy quick-cooking grains in the time that it takes to thaw and reheat them in your microwave—all for a huge savings.

Grain (1 cup dry)	Water (cups)	Cooking Time (min.)	Yield (cups)
Basmati rice	2	20	3
Brown rice	2	60	3
Bulgur	2	20	2½
Long-grain white rice	2	20	3
Millet	3	45	3½
Pearl barley	3	75	3½
Quinoa	2	30	3
Wild rice	3	60	4

Per serving
Calories 229
Total fat 2.9 g.
Saturated fat 0.6 g.
Cholesterol 3 mg.
Sodium 169 mg.
Fiber 7.4 g.

Cost per serving

29¢

WINTER KASHA WITH POTATOES AND CARROTS

Kasha is a common name for hulled, crushed buckwheat kernels. Look for toasted buckwheat, which has a wonderful nutty flavor, in supermarkets and natural food stores.

1 teaspoon olive oil
1 cup frozen chopped onions
1 cup frozen sliced carrots
2 small red potatoes, diced
1 cup kasha
2 cups frozen defatted Chicken Stock
     (page 61), thawed
½ teaspoon ground black pepper
¼ teaspoon salt
¼ cup minced fresh parsley

❋ Coat a medium saucepan with no-stick spray and place over medium-high heat until hot. Add the oil, onions, and carrots. Cook, stirring, for 5 minutes, or until the onions are soft but not browned. Add the potatoes and kasha. Cook, stirring, for 3 minutes.

❋ Add the stock. Bring to a boil. Reduce the heat to medium, cover, and cook for 15 minutes, or until the potatoes are soft and the kasha is tender. Add the pepper, salt, and parsley. Mix well.

ASIAN-STYLE WILD RICE SAUTÉ

Wild rice and white rice mix in this easy stove-top side dish. Wild rice, which is really a wild grass, is harvested from lakes in Canada and the northern United States. Save money by buying it in bulk in natural food stores.

½ cup frozen chopped onions
⅓ cup chopped celery
⅓ cup frozen whole kernel corn
⅓ cup frozen chopped sweet red peppers
1 cup wild rice
¼ cup apple juice
5 cups frozen defatted Chicken Stock (page 61), thawed
1 cup long-grain white rice
2 tablespoons chopped fresh parsley
1 teaspoon reduced-sodium soy sauce
½ teaspoon dark sesame oil
¼ teaspoon crushed red-pepper flakes

✳ Coat a Dutch oven with no-stick spray and place over medium-high heat until hot. Add the onions, celery, corn, and red peppers. Cook, stirring, for 5 minutes, or until the onions are soft but not browned. Add the wild rice and apple juice. Cook, stirring, for 5 minutes, or until the liquid evaporates.

✳ Add the stock. Bring to a boil. Reduce the heat to medium, cover, and cook for 10 minutes. Add the white rice and stir. Cover and cook for 30 minutes, or until the wild rice is tender. Add the parsley, soy sauce, oil, and red-pepper flakes. Stir well.

Makes 8 servings

Per serving
Calories 200
Total fat 1.1 g.
Saturated fat 0.2 g.
Cholesterol 4 mg.
Sodium 38 mg.
Fiber 2 g.

Cost per serving

30¢

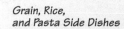

Per serving
Calories 96
Total fat 0.4 g.
Saturated fat 0.1 g.
Cholesterol 1 mg.
Sodium 136 mg.
Fiber 0.4 g.

Cost per serving

27¢

HERBED BASMATI RICE

This quick-cooking rice medley fills out a meal of roasted chicken or pork for less than 30¢ a serving.

½ cup basmati rice
3 cloves garlic, minced
1 cup frozen defatted Chicken Stock (page 61), thawed
¼ cup frozen chopped onions, minced
¼ cup frozen whole kernel corn
¼ cup frozen peas
1 teaspoon dried basil
½ teaspoon dried oregano
¼ teaspoon salt
2 tablespoons minced fresh parsley

FROM the Freezer

❋ Coat a large saucepan with no-stick spray and place over medium-high heat until hot. Add the rice and garlic. Cook, stirring, for 2 minutes. Add the stock, onions, corn, peas, basil, oregano, and salt. Bring to a boil. Reduce the heat to low, cover, and cook for 20 minutes, or until the rice is tender. Stir in the parsley.

BEST BUY ON BASMATI

Basmati is an especially tasty type of long-grain rice. Most basmati rice is imported from India, although some American companies grow it to meet increased demand. Basmati cooks to a chewy, but not sticky, softness. It's prized in Indian dishes and stir-fries. Asian markets sell it in bulk for less than 39¢ a pound, about 20¢ cheaper than the equivalent of boxed supermarket rice.

The secret to success with basmati rice is to rinse it prior to cooking to remove its natural white starch coating. Place the rice in a bowl and fill with cold water. Stir the rice with your hands to rinse it. Pour off the water and repeat until the rinse water is clear.

BUTTERNUT SQUASH RISOTTO

You save about $3.50 a serving over the restaurant version of this traditional Italian comfort food. Since Arborio rice creates its own creamy sauce, risotto freezes very well. If it separates when thawed, puree one-third in the blender and stir back in when reheating.

1 large butternut squash
5 cups frozen defatted Chicken Stock
   (page 61), thawed
1 tablespoon olive oil
3 cloves garlic, minced
2 cups Arborio rice
¼ cup chopped fresh parsley
¼ cup grated Parmesan cheese
¼ teaspoon salt
¼ teaspoon ground black pepper

❀ Pierce the squash several times with a sharp knife. Place it on a paper towel in the microwave. Microwave on high power for 5 minutes. Halve the squash lengthwise. Remove the seeds and discard. Place the halves back on the paper towel and microwave for 5 minutes, or until tender. Let cool slightly, then scoop the flesh into a bowl; discard the shells.

❀ Place the stock in a medium saucepan and bring to a simmer over medium-high heat. Reduce the heat to low; keep the stock warm.

❀ Coat a Dutch oven with no-stick spray and place over medium-high heat until hot. Add the oil and garlic. Cook, stirring, for 1 minute. Add the rice. Cook, stirring, for 5 minutes.

❀ Add the squash and ½ cup of the stock. Cook, stirring continuously, until the liquid has completely evaporated. Add ½ cup of the remaining stock. Cook, stirring, until the liquid has been absorbed.

❀ Continue to add stock, ½ cup at a time. After each addition, stir until the liquid is absorbed. Cook for a total of 25 minutes, or until the rice is creamy and tender.

❀ Stir in the parsley, Parmesan, salt, and pepper.

Makes 6 servings

Per serving
Calories 370
Total fat 3.9 g.
Saturated fat 1.2 g.
Cholesterol 8 mg.
Sodium 176 mg.
Fiber 4.1 g.

Cost per serving

70¢

KITCHEN TIP

To freeze, pack the cooled cooked risotto in a freezer-quality plastic container. To use, thaw overnight in the refrigerator. Cover and microwave on high power for 5 to 8 minutes, or until hot.

Grain, Rice, and Pasta Side Dishes

CURRIED SKILLET RICE

Indian curries look pretty with this pale yellow rice dish.

Makes 8 servings

Per serving
Calories 159
Total fat 1.3 g.
Saturated fat 0.2 g.
Cholesterol 3 mg.
Sodium 140 mg.
Fiber 0.8 g.

Cost per serving

37¢

1 teaspoon oil
1 cup frozen chopped onions
2 cloves garlic, minced
1½ cups basmati rice
1 teaspoon curry powder
4 cups frozen defatted Chicken Stock (page 61), thawed
1 tablespoon sugar
½ teaspoon salt

✻ Coat a large no-stick skillet with no-stick spray and place over medium-high heat until hot. Add the oil, onions, and garlic. Cook, stirring, for 5 minutes, or until the onions are soft but not browned. Add the rice and curry powder. Cook, stirring, for 2 minutes, or until fragrant. Add the stock, sugar, and salt. Bring to a boil. Reduce the heat to low, cover, and cook for 20 minutes, or until the rice is tender.

SPANISH RICE SALAD

If you add leftover cooked seafood, such as shrimp or fish, you can transform this side dish into a meal for no more than 50¢ a serving.

Makes 6 servings

Per serving
Calories 190
Total fat 5.1 g.
Saturated fat 0.8 g.
Cholesterol 2 mg.
Sodium 99 mg.
Fiber 1.7 g.

Cost per serving

31¢

Rice

½ cup frozen chopped onions
2 cloves garlic, minced
2 cups frozen defatted Chicken Stock (page 61), thawed
1 cup long-grain white rice

Salad

¼ cup lemon juice
2 tablespoons olive oil
1 cup halved cherry tomatoes
½ cup frozen chopped sweet red or yellow peppers
½ cup frozen chopped green peppers
⅓ cup chopped fresh parsley
¼ cup frozen peas
¼ teaspoon salt
¼ teaspoon ground black pepper

- *To make the rice:* Coat a 10″ no-stick skillet with no-stick spray and place over medium-high heat until hot. Add the onions and garlic. Cook, stirring, for 5 minutes, or until the onions are soft but not browned.

- Add the stock and rice. Bring to a boil. Reduce the heat to low, cover, and cook for 12 to 15 minutes, or until the rice is tender. Fluff the rice and spread out on a large plate. Refrigerate for 15 to 20 minutes, or until cool.

- *To make the salad:* In a large bowl, combine the lemon juice and oil. Mix well. Add the tomatoes, red or yellow peppers, green peppers, parsley, peas, salt, and black pepper. Add the rice. Toss well.

TOASTED RICE PILAF

Toasting the rice before steaming adds a nutty flavor to the pilaf. This easy side dish enhances an entrée for only 16¢ a serving, less than half of what you'd pay for commercial boxed pilaf mixes.

2 teaspoons olive oil
1 cup long-grain white rice
1 teaspoon minced garlic
½ teaspoon ground cumin
2 cups frozen defatted Chicken Stock
   (page 61), thawed
¼ teaspoon salt
¼ cup minced fresh parsley

- Coat a 10″ no-stick skillet with no-stick spray and place over medium-high heat until hot. Add the oil, rice, garlic, and cumin. Cook, stirring, for 3 to 5 minutes, or until fragrant.

- Add the stock and salt. Bring to a boil. Reduce the heat to low, cover, and cook for 20 minutes, or until the rice is tender. Sprinkle with the parsley.

Makes 4 servings

Per serving
Calories 216
Total fat 2.9 g.
Saturated fat 0.5 g.
Cholesterol 3 mg.
Sodium 140 mg.
Fiber 1 g.

Cost per serving

16¢

KITCHEN TIP

To freeze, pack the cooled cooked pilaf in a freezer-quality plastic bag. Flatten the bag. To use, thaw overnight in the refrigerator. Place in a 10″ no-stick skillet, cover, and cook, stirring occasionally, over low heat for 5 to 8 minutes, or until hot.

Grain, Rice, and Pasta Side Dishes

Per serving
Calories 276
Total fat 2.1 g.
Saturated fat 0.4 g.
Cholesterol 3 mg.
Sodium 66 mg.
Fiber 5.4 g.

Cost per serving

57¢

RICE AND CARROT SLAW

A tart ginger dressing adds zest to this side salad. It's a 5-minute dish if you have extra cooked rice stashed in the freezer. Just thaw the rice in the microwave and toss with the cooked vegetables.

- 2 cups frozen defatted Chicken Stock (page 61), thawed
- 1 cup long-grain white rice
- 2 cups shredded carrots
- 1 cup frozen peas
- ½ cup frozen broccoli florets
- ½ cup diced sweet red peppers
- 1 tablespoon lemon juice
- 1 tablespoon balsamic vinegar
- 1 teaspoon dark sesame oil
- 2 cloves garlic, minced
- ½ teaspoon minced fresh ginger
- ¼ cup chopped fresh cilantro

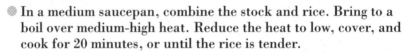

❋ In a medium saucepan, combine the stock and rice. Bring to a boil over medium-high heat. Reduce the heat to low, cover, and cook for 20 minutes, or until the rice is tender.

❋ In a 10″ no-stick skillet, combine the carrots, peas, broccoli, peppers, and lemon juice. Cover and cook over medium-high heat for 3 minutes, or until the broccoli is bright green. Transfer to a large bowl.

❋ To the bowl, add the vinegar, oil, garlic, ginger, and cilantro. Stir well. Add the rice and toss to coat.

CRANBERRY CASSEROLE STUFFING

Cranberries freeze very well, so you'll want to stock up during the fall, when the bright red fruit is abundant and cheap. To save time and effort during hectic holiday meal preparation, make this casserole stuffing up to 2 months ahead and store in the freezer.

  1 **teaspoon olive oil**
¼ **cup white wine or apple juice**
  1 **cup chopped onions**
½ **cup chopped celery**
  1 **clove garlic, minced**
  4 **cups soft bread crumbs**
¼ **cup cranberries**
  1 **teaspoon honey or sugar**
  1 **teaspoon dried thyme**
½ **teaspoon dried sage**
¼ **teaspoon ground black pepper**

❋ Preheat the oven to 400°F. Coat a 12″ × 8″ baking dish with no-stick spray.

❋ Coat a 10″ no-stick skillet with no-stick spray and place over medium-high heat until hot. Add the oil and wine or apple juice. Bring to a boil. Add the onions, celery, and garlic. Cook, stirring, for 5 minutes, or until the vegetables are soft but not browned.

❋ In a large bowl, combine the bread crumbs, cranberries, honey or sugar, thyme, sage, and pepper. Mix well. Add the onion mixture and mix well.

❋ Transfer to the prepared baking dish. Cover and bake for 20 minutes, or until heated through.

Makes 8 servings

Per serving
Calories 84
Total fat 1.6 g.
Saturated fat 0.4 g.
Cholesterol 1 mg.
Sodium 123 mg.
Fiber 1 g.

Cost per serving

18¢

KITCHEN TIP

To freeze, cool the cooked casserole. Wrap the baking dish in freezer-quality plastic wrap, then in freezer-quality foil. To use, thaw overnight in the refrigerator. Remove the foil and plastic wrap; discard the plastic wrap. Cover with the foil and bake at 350°F for 15 to 20 minutes, or until hot.

Grain, Rice, and Pasta Side Dishes

KITCHEN TIP

To freeze, cool the cooked casserole. Wrap the baking dish in freezer-quality plastic wrap, then in freezer-quality foil. To use, thaw overnight in the refrigerator. Remove the foil and plastic wrap; discard the plastic wrap. Cover with the foil and bake at 350°F for 15 to 20 minutes, or until hot.

THANKSGIVING BREAD STUFFING

This traditional New England stuffing can be a welcome side dish all year round for only 9¢ a serving.

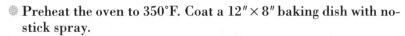

4 cups cubed reduced-sodium whole-wheat bread
1 teaspoon olive oil
4 stalks celery, minced
1 cup chopped onions
2 teaspoons poultry seasoning
½ teaspoon dried thyme
¼ teaspoon ground black pepper
1½ cups frozen defatted Chicken Stock (page 61), thawed

✤ Preheat the oven to 350°F. Coat a 12″ × 8″ baking dish with no-stick spray.

✤ Coat a large baking sheet with no-stick spray. Place the bread on the baking sheet and mist with no-stick spray. Bake for 15 minutes, or until golden brown. Transfer to a large bowl.

✤ Coat a 10″ no-stick skillet with no-stick spray and place over medium-high heat until hot. Add the oil, celery, onions, poultry seasoning, thyme, and pepper. Cook, stirring, for 5 minutes, or until the onions are soft but not browned.

✤ Add the stock to the bread cubes in the bowl. Toss to mix. Add the onion mixture and mix well. Transfer to the prepared baking dish. Bake for 30 minutes, or until golden brown.

Couscous with Currants

Couscous, a tiny grain-like pasta, is a favorite in Middle Eastern cooking. Even though couscous cooks quickly, it's still convenient to have this ready-to-serve side dish in the freezer for times when you need dinner on the table faster than fast.

2   cups frozen defatted Chicken Stock (page 61), thawed
½   cup currants
½   cup chopped scallions
¼   teaspoon salt
¼   teaspoon ground black pepper
1   cup couscous
2   tablespoons minced fresh parsley

❋ In a medium saucepan, combine the stock, currants, scallions, salt, and pepper. Bring to a boil over medium-high heat. Stir in the couscous. Remove from the heat. Cover and let stand for 10 minutes, or until the liquid is absorbed. Stir in the parsley.

Explore the World of Pasta

If you're feeling adventurous, substitute a new kind of pasta in your favorite recipe this week. In Italy, there are as many types of pasta as there are sauces to put over them. Flat noodles are common in the north (to allow the cream sauces to stick), while tubular shapes predominate in the south, where tomato sauce is king.

Most supermarkets carry a wide variety of Italian pastas, which can be mixed and matched with favorite pasta dishes in this chapter. Here are a few that are especially useful.

❋ *Conchigliette:* tiny shells

❋ *Farfalle:* bow-ties

❋ *Linguine:* long flat noodles

❋ *Penne:* tubular, sometimes ridged

❋ *Rotini:* corkscrews

❋ *Vermicelli:* a thinner version of spaghetti

Makes 4 servings

Per serving
Calories 194
Total fat 0.6 g.
Saturated fat 0.1 g.
Cholesterol 3 mg.
Sodium 144 mg.
Fiber 3.8 g.

Cost per serving

63¢

Kitchen Tip

To freeze, pack the cooled cooked couscous in a freezer-quality plastic bag. To use, thaw overnight in the refrigerator. Place in a 10" no-stick skillet. Cover and cook, stirring frequently, over low heat for 3 to 5 minutes, or until hot.

Farfalle with Feta

This Greek side dish also makes a delightful light main dish for 4. Far-falle noodles are also known as bow-ties or butterflies.

Makes 6 servings

Per serving
Calories 191
Total fat 4.6 g.
Saturated fat 2 g.
Cholesterol 13 mg.
Sodium 165 mg.
Fiber 2.2 g.

Cost per serving

39¢

 8 ounces farfalle pasta
 1 teaspoon olive oil
 ½ cup frozen broccoli florets, chopped
 ½ cup frozen chopped sweet red peppers
 ½ cup chopped tomatoes
 2 cloves garlic, minced
 ¼ teaspoon dried thyme
 ¾ cup crumbled feta cheese
 3 tablespoons chopped fresh parsley
 ¼ teaspoon ground black pepper

✤ Cook the pasta in a large pot of boiling water according to the package directions. Drain well and return to the pot.

✤ Coat a 10″ no-stick skillet with no-stick spray and place over medium-high heat until hot. Add the oil, broccoli, red peppers, tomatoes, garlic, and thyme. Cook, stirring, for 5 to 8 minutes, or until soft. Add the feta. Cook, stirring, for 2 minutes, or until the cheese begins to melt.

✤ Pour the sauce over the pasta. Add the parsley and black pepper. Toss well.

Pointers for Perfect Pasta

Follow these chefs' tips for perfectly cooked pasta every time.

✤ Use a big pot and plenty of water to boil your pasta so that it can move freely and not stick together. Most chefs recommend 4 quarts of water for each pound of pasta.

✤ If you are on a sodium-restricted eating plan, add a pinch of salt to the sauce, not the pasta water. If sodium isn't a health concern, you may choose—as most Italian chefs do—to add a little salt to the cooking water.

✤ Forgo the oil in the pasta cooking water. It adds unnecessary fat, and can make the pasta greasy enough to prevent it from absorbing the flavor of the sauce.

CREAMY SPINACH-PARMESAN ORZO

A creamy low-fat sauce coats the tiny rice-shaped pasta called orzo in this comforting side dish. Individually flash-frozen spinach leaves and sweet red peppers make this dish convenient because you don't have to thaw them before measuring or cooking.

1 cup orzo pasta
1 teaspoon olive oil
2 cloves garlic, chopped
2 cups packed frozen spinach, chopped
½ cup frozen chopped sweet red peppers
½ cup low-fat ricotta cheese
¼ cup grated Parmesan cheese
½ teaspoon ground black pepper

❋ Cook the pasta in a large saucepan of boiling water according to the package directions. Drain well. Place in a medium bowl.

❋ Coat a 10″ no-stick skillet with no-stick spray and place over medium-high heat until hot. Add the oil and garlic. Cook, stirring, for 2 minutes. Add the spinach and red peppers. Cook, stirring, for 8 to 10 minutes, or until the liquid from the spinach has evaporated.

❋ In a blender or food processor, combine the ricotta, Parmesan, and black pepper. Process until smooth. Pour over the pasta. Add the cooked vegetables. Toss well.

Makes 6 servings

Per serving
Calories 201
Total fat 4.3 g.
Saturated fat 1 g.
Cholesterol 13 mg.
Sodium 250 mg.
Fiber 2.4 g.

Cost per serving

52¢

Per serving
Calories 284
Total fat 3.6 g.
Saturated fat 1 g.
Cholesterol 4 mg.
Sodium 284 mg.
Fiber 8.2 g.

Cost per serving

49¢

PASTA AND NAVY BEANS WITH SAGE

If you freeze your cooked beans in 1-cup packages, they'll thaw quickly in the microwave for recipes such as this. You can add slivered chicken, pork, or lean beef to this side dish for a fast light supper.

1 teaspoon olive oil
1 cup frozen chopped onions
1 cup frozen sliced carrots
4 cloves garlic, minced
2 cups chopped tomatoes
1 cup frozen cooked navy beans (page 177), thawed
½ cup orzo pasta
1 cup frozen defatted Chicken Stock (page 61), thawed
1 teaspoon dried sage
½ teaspoon dried thyme
¼ teaspoon ground black pepper
¼ teaspoon salt
¼ cup dry bread crumbs
2 tablespoons grated Parmesan cheese

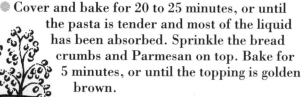

* Preheat the oven to 425°F.

* Coat a Dutch oven with no-stick spray and place over medium-high heat until hot. Add the oil, onions, carrots, and garlic. Cook, stirring, for 5 to 8 minutes, or until the onions are soft but not browned.

* Add the tomatoes, beans, and pasta. Cook, stirring, for 3 to 5 minutes, or until the tomatoes soften. Add the stock, sage, thyme, pepper, and salt. Bring to a boil.

* Cover and bake for 20 to 25 minutes, or until the pasta is tender and most of the liquid has been absorbed. Sprinkle the bread crumbs and Parmesan on top. Bake for 5 minutes, or until the topping is golden brown.

WARM SALAD-BOWL PASTA

Try a variety of freezer vegetables for this easy year-round salad. The dish is tastiest if served within 4 hours of preparation.

  4 ounces elbow macaroni
½ cup frozen broccoli florets
½ cup frozen whole kernel corn
½ cup frozen green beans
  1 tablespoon water
½ cup halved cherry tomatoes
  2 tablespoons olive oil
  2 tablespoons minced fresh parsley or basil
  2 tablespoons balsamic vinegar
  1 teaspoon sugar
  1 teaspoon Dijon mustard
¼ teaspoon salt
¼ teaspoon ground black pepper
  2 tablespoons grated Parmesan

✻ Cook the macaroni in a large pot of boiling water according to the package directions. Drain well. Place in a large bowl.

✻ In a 10″ no-stick skillet, combine the broccoli, corn, beans, and water. Cover and cook over medium-high heat for 2 minutes, or until the broccoli is bright green. Drain in a colander and rinse under cold water. Pat dry with paper towels. Add to the bowl with the macaroni. Add the tomatoes, oil, parsley or basil, vinegar, sugar, mustard, salt, and pepper. Toss well. Sprinkle with the Parmesan.

Makes 6 servings

Per serving
Calories 147
Total fat 5.6 g.
Saturated fat 1.1 g.
Cholesterol 1 mg.
Sodium 157 mg.
Fiber 1.9 g.

Cost per serving
21¢

KITCHEN TIP

Get full value from any bunch of herbs you purchase. After using the amount you need for a recipe, chop and freeze the remaining leaves in a small freezer-quality plastic bag. Parsley, basil, and other herbs will remain green and flavorful. Use the herbs while still frozen.

Grain, Rice, and Pasta Side Dishes

Yeast Breads, Muffins, and Quick Breads

W hen something wonderful is in the oven, the family flocks to the kitchen, drawn by the aroma of baking. To economize on time, many cooks stock their freezers with favorite baked goods. Since freezing doesn't noticeably change bread's texture or taste, it is the best way to preserve the just-made quality of breads, muffins, and other oven treats.

Freezing is also the best resource to have baked goods on hand for the least cost. A loaf of whole-grain bread from your own oven costs $1 to $2.50, while you pay $3 or $4 for store-bought breads. Muffins save you even more—you pocket close to $1 on each muffin on some varieties when you make your own.

Use the recipes in this chapter to stock your freezer with homemade baked goods that take only minutes of hands-on work yet fill the kitchen with that wonderful homey aroma.

229

Per serving
Calories 157
Total fat 1.8 g.
Saturated fat 0.6 g.
Cholesterol 2 mg.
Sodium 42 mg.
Fiber 1.5 g.

Cost per serving

7¢

KITCHEN TIP

To freeze, wrap the cooled focaccia in freezer-quality plastic wrap, then in freezer-quality foil. To use, thaw overnight in the refrigerator or thaw in the microwave. To reheat, remove the foil and plastic wrap; discard the plastic wrap. Rewrap in the foil and bake at 425°F for 15 minutes, or until hot.

ONION-CHEESE FOCACCIA

This Italian flatbread is a delightful addition to a company meal. Focaccia freezes well and tastes freshly baked even after thawing and reheating.

1	cup warm water (about 110°F)
¼	cup sugar
1	tablespoon or 1 package active dry yeast
⅓	cup chopped onions
2	teaspoons olive oil
3	cups all-purpose flour
¼	cup grated Parmesan cheese
2	tablespoons minced garlic
½	teaspoon ground red pepper
2	tablespoons yellow cornmeal
1	teaspoon dried thyme

✺ In a large bowl, combine the water, sugar, and yeast. Stir well. Let stand in a warm place for 5 minutes, or until foamy.

✺ Place a 10″ no-stick skillet over medium-high heat until hot. Add the onions and 1 teaspoon of the oil. Cook, stirring, for 3 minutes, or until slightly softened. Let cool. Add to the yeast mixture. Stir well.

✺ Stir in 1½ cups of the flour. Cover and set in a warm place for 30 minutes, or until doubled in size. Stir in the Parmesan, garlic, red pepper, and about 1 cup of the remaining flour to make a kneadable dough. Turn out onto a lightly floured work surface. Knead, adding up to ½ cup more flour, for 10 minutes, or until smooth and elastic.

✺ Coat a large baking sheet with no-stick spray and sprinkle with the cornmeal.

✺ Roll the dough into a 12″ × 9″ rectangle. Place on the prepared baking sheet. Cover and set in a warm place for 30 minutes, or until doubled in size.

✺ Preheat the oven to 425°F.

✺ With your fingertips, press ¼″-deep indentations over the surface of the dough. Brush with the remaining 1 teaspoon oil and sprinkle with the thyme. Bake for 20 minutes, or until golden brown.

ROSEMARY-ONION WHEAT BAGUETTES

This fragrant bread is in the shape of small, long narrow loaves, enough to serve 2 people each, and just perfect alongside soup and salad for a hearty supper. Each baguette saves you 22¢ over store-bought.

1 cup chopped onions
1 teaspoon olive oil
1 teaspoon chopped fresh rosemary
2 cups warm water (about 110°F)
½ cup honey or brown sugar
1 tablespoon or 1 package active dry yeast
4 cups whole-wheat flour
1 teaspoon dried thyme
¼ teaspoon salt
1 egg white, lightly beaten

❋ Place a 10" no-stick skillet over medium-high heat until hot. Add the onions, oil, and rosemary. Cook, stirring, for 5 minutes, or until the onions are soft but not browned. Let cool.

❋ In a large bowl, combine the water, honey or brown sugar, and yeast. Stir well. Let stand in a warm place for 5 minutes, or until foamy.

❋ Stir 3 cups of the flour into the bowl. Cover and let stand in a warm place for 30 minutes, or until doubled in size.

❋ Stir in the onion mixture, thyme, salt, and about ½ cup of the remaining flour to make a kneadable dough. Turn the dough out onto a lightly floured surface. Knead, adding up to ½ cup more flour, for 10 minutes, or until smooth and elastic.

❋ Coat another large bowl with no-stick spray. Add the dough and turn to coat all sides. Cover and set in a warm place for 1 hour, or until doubled in size.

❋ Coat 2 large baking sheets with no-stick spray. Divide the dough into 4 equal balls. On a lightly floured surface, roll each ball into a 12" rope. Place on the prepared baking sheets. Cut 3 slits in the top of each baguette. Brush with the egg white. Let stand for 15 minutes.

❋ Preheat the oven to 350°F. Bake for 35 minutes, or until golden brown.

Makes 8 servings

Per serving
Calories 286
Total fat 1.8 g.
Saturated fat 0.3 g.
Cholesterol 0 mg.
Sodium 83 mg.
Fiber 7.7 g.

Cost per serving

13¢

KITCHEN TIP

To freeze, wrap the cooled baguettes in freezer-quality plastic wrap, then in freezer-quality foil. To use, thaw overnight in the refrigerator. To reheat, remove the foil and plastic wrap; discard the plastic wrap. Rewrap in the foil. Bake at 425°F for 15 minutes, or until hot.

Yeast Breads, Muffins, and Quick Breads

Per slice
Calories 106
Total fat 1.8 g.
Saturated fat 0.9 g.
Cholesterol 4 mg.
Sodium 76 mg.
Fiber 1.2 g.

Cost per serving

10¢

ITALIAN PIZZA BREAD

Try this loaf version of traditional pizza. Savory herb-scented bread is formed into a long strip, filled with pizza ingredients, then rolled into a baguette. When you find mozzarella at a good price, buy extra to shred and freeze for up to 4 months.

Filling

½ cup frozen chopped onions
½ cup frozen chopped sweet red peppers
½ cup chopped mushrooms
1 teaspoon olive oil
1 cup reduced-sodium pizza sauce
½ teaspoon Italian herb seasoning

Dough

1 cup warm water (about 110°F)
1 tablespoon or 1 package active dry yeast
1 tablespoon sugar
2½ cups all-purpose flour
¼ teaspoon salt
1 cup shredded low-fat mozzarella cheese

* ***To make the filling:*** Place a 10″ no-stick skillet over medium-high heat until hot. Add the onions, peppers, mushrooms, and oil. Cook, stirring, for 5 minutes, or until the onions are soft but not browned. Add the pizza sauce and Italian herb seasoning. Bring to a simmer. Remove from the heat and let cool.

* ***To make the dough:*** In a large bowl, combine the water, yeast, and sugar. Stir well. Let stand in a warm place for 5 minutes, or until foamy.

* Stir in 1 cup of the flour. Cover and set in a warm place for 30 minutes, or until doubled in size.

* Stir in the salt and about 1 cup of the remaining flour to make a kneadable dough. Turn the dough out onto a lightly floured surface. Knead, adding up to ½ cup more flour, for 10 minutes, or until smooth and elastic.

* Coat another large bowl with no-stick spray. Add the dough and turn to coat all sides. Cover and set in a warm place for 1 hour, or until doubled in size.

* Coat a large baking sheet with no-stick spray. On a lightly floured surface, roll the dough into a 14" × 10" rectangle. Spread with the filling, then sprinkle with the mozzarella. Roll up, pinching the ends and edges together to seal. Place, seam side down, on the prepared baking sheet. Let stand for 15 minutes.

* Preheat the oven to 350°F. Bake for 45 to 60 minutes, or until golden brown. Let cool before slicing.

FREEZING YEAST DOUGHS

Yeast bread dough keeps for one month in the freezer with no loss in quality, according to Shirley Corriher, author of *CookWise*. Follow these tips for freezing in flavor.

* Prepare the dough and let it rise once.

* Punch down, then divide into balls (the number of loaves in the recipe).

* Pack each dough ball in a gallon-size freezer-quality plastic bag, flattening the balls and squeezing as much air from the bag as possible before sealing.

* Label the bags with the recipe name, recipe yield, and the date.

* Place the bags on the bottom shelf—the coldest part of the freezer—to fast-freeze the dough.

* To thaw, open the bag and remove the dough. Place in a medium bowl. Cover and let thaw in the refrigerator overnight.

* Bring to room temperature and continue with the recipe.

YEAST: THE LIVING LEAVENING

Active dry yeast—the familiar dehydrated granules—is dormant but still alive. When mixed with liquid that is approximately 110° to 115°F, the yeast comes back to life and starts to grow. This process is what makes yeast breads rise.

Packaged dry yeast has an expiration date for refrigerated storage, but it keeps practically forever in the freezer, says Amy Dacyczyn, thrift expert and author of *The Tightwad Gazette* series. Keep the yeast in a tightly sealed container to prevent moisture from rehydrating the yeast before you're ready to use it. Remove the amount you need from the freezer before starting a recipe and let it come to room temperature.

When you purchase bulk yeast from a natural food store, you save a whopping 64¢ per tablespoon. To use, just measure out a scant tablespoon to replace each package of dry yeast called for in a recipe. Standard American dry yeast packets contain 7 grams or 2¼ teaspoons of yeast, which is ¾ teaspoon shy of 1 tablespoon. All of the yeast dough recipes in this book work with either 1 tablespoon or 1 packet of dry yeast.

BUTTERMILK AND CHIVE DINNER ROLLS

Most yeast breads can be formed into a variety of shapes—including these dinner rolls. They freeze well, and leftover rolls are great for brown-bag sandwiches.

1¼ cups low-fat buttermilk
 3 tablespoons sugar
 1 tablespoon or 1 package active dry yeast
 4 cups all-purpose flour
¾ cup yellow cornmeal
 3 tablespoons oil
 2 tablespoons minced fresh chives
 1 egg white, lightly beaten

* Place the buttermilk in a small saucepan. Warm over medium heat (about 110°F). Pour into a large bowl. Add the sugar and yeast. Stir well. Let stand in a warm place for 5 minutes, or until foamy.

* Stir in 2 cups of the flour. Cover and set in a warm place for 30 minutes, or until doubled in size.

* Stir in the cornmeal, oil, and chives, and about 1½ cups of the remaining flour to make a kneadable dough. Turn the dough out onto a lightly floured surface. Knead, adding up to ½ cup more flour, for 10 minutes, or until smooth and elastic.

* Coat another large bowl with no-stick spray. Add the dough and turn to coat all sides. Cover and set in a warm place for 1 hour, or until doubled in size.

* Coat a large baking sheet with no-stick spray. Divide the dough into 18 balls. Roll each ball into a 6" rope. Form a tight circle with each rope; pinch the ends together to seal. Place the rolls on the prepared baking sheet, leaving about 2" space between rolls. Cover and set in a warm place for 10 minutes.

* Preheat the oven to 375°F. Brush the rolls with the egg white. Bake for 15 to 20 minutes, or until the rolls are golden brown.

Makes 18

Per roll
Calories 157
Total fat 2.9 g.
Saturated fat 0.3 g.
Cholesterol 1 mg.
Sodium 23 mg.
Fiber 1.4 g.

Cost per serving

8¢

KITCHEN TIP

To freeze, pack the cooled rolls in a freezer-quality plastic bag. To use, thaw overnight in the refrigerator. To reheat, wrap in foil and bake at 425°F for 15 minutes, or until hot.

Yeast Breads, Muffins, and Quick Breads

Per slice
Calories 264
Total fat 1.7 g.
Saturated fat 0.2 g.
Cholesterol 0 mg.
Sodium 6 mg.
Fiber 2.5 g.

Cost per serving

17¢

KITCHEN TIP

The cooled loaf can
be frozen whole or
sliced. To freeze,
wrap in freezer-
quality plastic
wrap, then in
freezer-quality foil.
To use the whole
loaf, thaw
overnight in the
refrigerator.
Remove the foil
and plastic wrap;
discard the plastic
wrap. Rewrap in
the foil. Bake at
350°F for
15 minutes, or until
hot. To use single
slices, remove from
the freezer and
place in a toaster
or toaster oven
until warm.

CINNAMON-RAISIN BREAD

*Spices permeate this breakfast bread, and a cinnamon-raisin filling
spirals through it. It's irresistible when toasted.*

Dough

1 cup apple juice
½ cup sugar
1 tablespoon or 1 package active dry yeast
1 tablespoon ground cinnamon
1 teaspoon pumpkin pie spice
4 cups all-purpose flour
1 tablespoon oil

Filling

1 cup raisins
¼ cup frozen apple juice concentrate, thawed
1 teaspoon ground cinnamon

* *To make the dough:* Place the apple juice in a small saucepan.
Warm over medium heat (about 110°F). Pour into a large bowl.
Add the sugar and yeast. Stir well. Let stand in a warm place for
5 minutes, or until foamy.

* Add the cinnamon, pumpkin pie spice, and 2½ cups of the flour.
Stir well. Cover and set in a warm place for 30 minutes, or until
doubled in size.

* Stir in the oil and about 1 cup of the remaining flour to make a
kneadable dough. Turn the dough out onto a lightly floured surface. Knead, adding up to ½ cup more flour, for 10 minutes, or
until smooth and elastic.

* Coat another large bowl with no-stick spray. Add the dough and
turn to coat all sides. Cover and set in a warm place for 1 hour,
or until doubled in size.

* *To make the filling:* In a microwaveable bowl, combine the
raisins, apple juice concentrate, and cinnamon. Microwave on
medium power for 5 minutes, or until thickened. Let cool.

* Preheat the oven to 350°F. Coat a 9″ × 5″ loaf pan with no-stick
spray.

- On a lightly floured surface, roll the dough into a 10″ × 8″ rectangle. Spread the filling over the dough, then roll up. Pinch the seams to seal. Place, seam side down, in the prepared pan. Let stand for 5 minutes.

- Bake for 45 minutes, or until golden brown. Remove from the pan and cool on a wire rack before slicing.

ANADAMA BREAD

This New England bread is Yankee thrifty at 10¢ a slice and hearty with honest ingredients like cornmeal and molasses. Soaking the cornmeal in boiling water gives the bread a chewy texture.

1   cup boiling water
½   cup yellow cornmeal
⅓   cup molasses
2   tablespoons oil
1   tablespoon active dry yeast
3   cups all-purpose flour
1   egg, lightly beaten
¼   teaspoon salt

- In a medium bowl, combine the water, cornmeal, molasses, and oil. Stir well. Let stand at room temperature for 15 minutes, or until cooled to warm (about 110°F).

- Stir in the yeast and 2 cups of the flour. Cover and set in a warm place for 30 minutes, or until doubled in size.

- Stir in the egg, salt, and about ½ cup of the remaining flour to make a kneadable dough. Turn the dough out onto a lightly floured surface. Knead, adding up to ½ cup more flour, for 10 minutes, or until smooth and elastic.

- Coat a large bowl with no-stick spray. Add the dough and turn to coat all sides. Cover and set in a warm place for 1 hour, or until doubled in size.

- Preheat the oven to 350°F. Coat a 9″ × 5″ loaf pan with no-stick spray.

- On a lightly floured surface, roll the dough into a 10″ × 8″ rectangle. Roll up. Pinch the seams to seal. Place, seam side down, in the prepared pan. Let stand for 15 minutes.

- Bake for 45 minutes, or until golden brown. Remove from the pan and cool on a wire rack before slicing.

Yeast Breads, Muffins, and Quick Breads

FIG-APPLE SWEET BREAD

This glazed party bread is studded with a variety of dried fruits and spices. You can substitute other dried fruits—raisins, dates, cranberries, or apricots—for the chopped figs and apples.

2   cups warm water (about 110°F)
½   cup honey or brown sugar
1   tablespoon or 1 package active dry yeast
4   cups all-purpose flour
⅓   cup chopped dried figs
⅓   cup chopped dried apples
1   tablespoon oil
1   teaspoon ground cinnamon
½   teaspoon ground allspice
¼   cup confectioners' sugar
2   tablespoons skim milk

✽ In a large bowl, combine the water, honey or brown sugar, and yeast. Stir well. Let stand in a warm place for 5 minutes, or until foamy.

✽ Stir in 3 cups of the flour. Cover and let stand in a warm place for 30 minutes, or until doubled in size.

✽ Stir in the figs, apples, oil, cinnamon, allspice, and about ½ cup of the remaining flour to make a kneadable dough. Turn the dough out onto a lightly floured surface.

✽ Knead, adding up to ½ cup more flour, for 10 minutes, or until smooth and elastic.

✽ Coat another large bowl with no-stick spray. Add the dough and turn to coat all sides. Cover and let stand in a warm place for 1 hour, or until doubled in size.

✽ Coat a 9" × 5" loaf pan with no-stick spray. On a lightly floured surface, roll the dough into a 10" × 8" rectangle. Roll up. Pinch the seams to seal. Place, seam side down, in the prepared pan. Let stand for 15 minutes.

✽ Preheat the oven to 350°F. Bake for 45 minutes, or until golden brown. Cool on a wire rack for 10 minutes. Remove the bread from the pan and cool on the wire rack.

✽ In a small bowl, combine the confectioners' sugar and milk. Drizzle over the loaf.

SPICY SOUTHWESTERN CORNBREAD

Monterey Jack cheese plus sweet peppers and jalapeño peppers give this staple southern bread a Southwestern flair. At only 24¢ a serving, you'll want to enjoy it often.

3   cups yellow cornmeal
¼   cup sugar
½   teaspoon salt
2   cups boiling water
3   tablespoons oil
1   cup skim milk
1   egg, lightly beaten
⅓   cup shredded low-fat Monterey Jack cheese
¼   cup frozen chopped sweet red peppers
¼   cup frozen chopped onions
1   tablespoon baking powder
1   teaspoon minced jalapeño peppers
     (wear plastic gloves when handling)

❋ Preheat the oven to 400°F. Coat a 13" × 9" baking dish with no-stick spray.

❋ In a large bowl, combine the cornmeal, sugar, and salt. Mix well. Add the water and oil. Mix well.

❋ Add the milk and egg. Mix well. Add the Monterey Jack, red peppers, onions, baking powder, and jalapeño peppers. Mix well. Pour into the prepared baking dish.

❋ Bake for 40 minutes, or until a toothpick inserted in the center of the cornbread comes out clean. Cool before cutting.

Makes 8 servings

Per serving
Calories 266
Total fat 7.5 g.
Saturated fat 0.8 g.
Cholesterol 28 mg.
Sodium 396 mg.
Fiber 3.5 g.

Cost per serving

24¢

KITCHEN TIP

To keep cornmeal fresh for up to 6 months, place in a freezer-quality plastic container or freezer-quality plastic bag and store in the freezer.

To freeze hot peppers, just seed and mince (wear plastic gloves when handling). Freeze on a baking sheet to keep loose, then package in freezer-quality plastic bags. Hot peppers can be frozen for up to 5 months.

Per slice
Calories 149
Total fat 3.1 g.
Saturated fat 0.4 g.
Cholesterol 14 mg.
Sodium 73 mg.
Fiber 1.2 g.

Cost per serving

16¢

KITCHEN TIP

The cooled loaf
can be frozen
whole or sliced. To
freeze, wrap in
freezer-quality
plastic wrap, then
in freezer-quality
foil. To use the
whole loaf, thaw
overnight in the
refrigerator.
Remove the foil
and plastic wrap;
discard the plastic
wrap. Rewrap in
the foil. Bake at
350°F for 15 min-
utes, or until hot.
To use single
slices, remove from
the freezer and
place in a toaster
or toaster oven
until warm.

GERMAN STOLLEN

*Traditional brunch fare, this sweet bread is studded with raisins and al-
monds, then glazed with sugar. Instead of a cup of butter, we use cottage
cheese to get a light texture yet keep the fat grams low.*

¾ cup skim milk
¼ cup packed brown sugar
1 tablespoon or 1 package active dry yeast
1 cup low-fat cottage cheese
2½ cups all-purpose flour
¾ cup raisins
1 egg, lightly beaten
2 tablespoons finely chopped almonds
2 tablespoons oil
2 teaspoons vanilla
1 teaspoon ground cinnamon
½ teaspoon ground cardamom
1 teaspoon reduced-calorie butter, melted
1 teaspoon sugar

⬧ Place the milk in a small saucepan. Warm over medium heat (about
110°F). Pour into a large bowl. Add the brown sugar and yeast.
Stir well. Let stand in a warm place for 5 minutes, or until foamy.

⬧ Stir in the cottage cheese and 1 cup of the flour. Cover and set in
a warm place for 30 minutes, or until doubled in size. Stir in the
raisins, egg, almonds, oil, vanilla, cinnamon, cardamom, and
1 cup of the remaining flour to make a kneadable dough.

⬧ Turn the dough out onto a lightly floured surface. Knead,
adding up to ½ cup more flour, for 10 minutes, or until smooth
and elastic.

⬧ Coat another large bowl with no-stick spray. Add the dough and
turn to coat all sides. Cover and let stand in a warm place for
1 hour, or until doubled in size.

⬧ Preheat the oven to 350°F. Coat a large baking sheet with no-
stick spray.

⬧ Pat the dough into a round loaf, then place on the prepared
baking sheet. Brush with the butter, then sprinkle with the
sugar. Bake for 45 to 55 minutes, or until golden brown. Remove
from the pan and cool on a wire rack. To serve, cut the loaf into
16 wedges.

CRANBERRY-WALNUT DATE BREAD

Fresh cranberries are widely available in the fall and, because of their firm shells, keep very well in the freezer. To freeze, just slit the plastic bag in which the cranberries are packed. Squeeze out as much air as possible and pack that bag into a freezer-quality plastic bag. Squeeze out the air and seal tightly. When you need them, it's easy to measure out just the amount of cranberries you need and store the rest back in the freezer.

½ cup fresh or frozen cranberries
½ cup chopped pitted dates
½ cup low-fat buttermilk
3 tablespoons packed brown sugar
1 egg
2 tablespoons unsweetened applesauce
2 tablespoons oil
1½ cups all-purpose flour
1 tablespoon toasted chopped walnuts
1 teaspoon baking powder
1 teaspoon baking soda

❋ Preheat the oven to 400°F. Coat a 9" × 5" loaf pan with no-stick spray. Dust the pan with flour and shake out the excess.

❋ In a medium bowl, combine the cranberries, dates, buttermilk, brown sugar, egg, applesauce, and oil. Mix well.

❋ In a large bowl, combine the flour, walnuts, baking powder, and baking soda. Mix well. Add the cranberry mixture. Stir until just blended. Pour into the prepared loaf pan.

❋ Bake for 25 minutes. Reduce the heat to 300°F. Bake for 15 minutes, or until a toothpick inserted in the center of the loaf comes out clean. Cool on a wire rack for 10 minutes. Remove the bread from the pan and cool on the wire rack.

PACK BREADS RIGHT

To freeze bread, cover each loaf tightly with freezer-quality plastic wrap, then wrap in freezer-quality foil or place in a freezer-quality plastic bag. Squeeze air from the bag before sealing. Label and date each package. For convenience, slice breads before freezing. Pull out the number of slices you need. Let them thaw at room temperature or reheat them in the toaster or toaster oven.

Makes 12 slices

Per slice
Calories 128
Total fat 3.4 g.
Saturated fat 0.4 g.
Cholesterol 18 mg.
Sodium 164 mg.
Fiber 1.3 g.

Cost per serving

19¢

KITCHEN TIP

The cooled loaf can be frozen whole or sliced. To freeze the cooled bread, wrap in freezer-quality plastic wrap, then in freezer-quality foil. To use the whole loaf, thaw overnight in the refrigerator. To reheat, place the foil-wrapped loaf in a 400°F oven for 10 minutes, or until hot. To use single slices, remove from the freezer and place in a toaster oven until warm.

Yeast Breads, Muffins, and Quick Breads

CHOCOLATE CHIP MUFFINS

*Even though they'd charm any chocolate lover, these muffins have only
4 grams of fat each. Plus, they cost 88¢ less each than their supermarket
counterparts.*

¾ cup low-fat buttermilk
½ cup packed brown sugar
1 egg
1 egg white
2 tablespoons unsweetened applesauce
2 tablespoons oil
1½ cups all-purpose flour
¼ cup cocoa powder
¼ cup semisweet chocolate chips
1½ teaspoons baking powder
½ teaspoon baking soda
¼ teaspoon salt

❀ Preheat the oven to 400°F. Coat a 12-cup no-stick muffin tin
with no-stick spray.

❀ In a medium bowl, combine the buttermilk, brown sugar, egg,
egg white, applesauce, and oil. Mix well.

❀ In a large bowl, combine the flour, cocoa, chocolate chips,
baking powder, baking soda, and salt. Mix well. Add the buttermilk mixture. Stir until just blended. Pour into the prepared
muffin tin, filling the cups three-fourths full.

❀ Bake for 12 to 15 minutes, or until a toothpick inserted in the
center of a muffin comes out clean.

CARROT-RAISIN MUFFINS

Makes 12 muffins

Per muffin
Calories 164
Total fat 3.4 g.
Saturated fat 0.4 g.
Cholesterol 18 mg.
Sodium 258 mg.
Fiber 1.7 g.

Cost per serving

14¢

Carrot-cake muffins are sweet enough for dessert. Freeze for youngsters' club meetings or treats anytime.

¾ cup unsweetened applesauce
⅓ cup packed brown sugar
2 egg whites
1 egg
2 tablespoons oil
1 teaspoon vanilla
2 cups all-purpose flour
1½ cups shredded carrots
½ cup raisins
1 tablespoon toasted chopped walnuts
2 teaspoons ground cinnamon
1 teaspoon baking powder
1 teaspoon baking soda
½ teaspoon salt

✽ Preheat the oven to 400°F. Coat a 12-cup no-stick muffin tin with no-stick spray.

✽ In a medium bowl, combine the applesauce, brown sugar, egg whites, egg, oil, and vanilla. Mix well.

✽ In a large bowl, combine the flour, carrots, raisins, walnuts, cinnamon, baking powder, baking soda, and salt. Mix well. Add the applesauce mixture. Stir until just blended. Pour into the prepared muffin tin, filling the cups three-fourths full.

✽ Bake for 15 to 18 minutes, or until a toothpick inserted in the center of a muffin comes out clean.

KITCHEN TIP

To freeze, place the cooled muffins on a tray. Put in the freezer for 1 hour, or until solid. Pack in a freezer-quality plastic bag. To use, thaw the number of muffins you need overnight in the refrigerator. Reheat at 400°F for 10 minutes, or until hot.

KITCHEN TIP

To freeze, place the cooled muffins on a tray. Put in the freezer for 1 hour, or until solid. Pack in a freezer-quality plastic bag. To use, thaw the number of muffins you need overnight in the refrigerator. Reheat at 400°F for 10 minutes, or until hot.

GINGER BRAN MUFFINS

These moist gingery muffins hide a healthy punch. Bran flakes contribute fiber and prune puree eliminates ½ cup of oil. That's a savings of more than 100 grams of fat in just one recipe!

¾ cup prune puree
½ cup low-fat buttermilk
⅓ cup molasses
⅓ cup packed brown sugar
¼ cup oil
1 egg white
1 egg
1 teaspoon vanilla
2 cups all-purpose flour
½ cup bran flakes
½ cup raisins
2 teaspoons ground cinnamon
2 teaspoons ground ginger
1 teaspoon baking powder
1 teaspoon baking soda
½ teaspoon ground cloves

✸ Preheat the oven to 400°F. Coat a 12-cup no-stick muffin tin with no-stick spray.

✸ In a medium bowl, combine the prune puree, buttermilk, molasses, brown sugar, oil, egg white, egg, and vanilla. Mix well.

✸ In a large bowl, combine the flour, bran flakes, raisins, cinnamon, ginger, baking powder, baking soda, and cloves. Mix well. Add the prune mixture. Stir until just blended. Pour into the prepared muffin tin, filling the cups three-fourths full.

✸ Bake for 20 minutes, or until a toothpick inserted in the center of a muffin comes out clean.

BROWN SUGAR–APPLE MUFFINS

Toasting the nuts brings out their flavor so that you can use less of this costly, high-fat ingredient. To toast, place the nuts in a dry skillet over medium heat. Toast, shaking the pan often, for 3 to 5 minutes, or until fragrant.

Topping

    1/4 cup all-purpose flour
    1/4 cup packed brown sugar
    1/2 teaspoon ground cinnamon
      1 tablespoon reduced-calorie butter, melted

Muffins

    1/2 cup chopped apples
    1/2 cup chopped pitted dates
    1/2 cup low-fat buttermilk
      3 tablespoons packed brown sugar
      1 egg
      2 tablespoons unsweetened applesauce
      2 tablespoons oil
    1 1/2 cups all-purpose flour
      1 tablespoon toasted chopped walnuts
      1 teaspoon baking powder
      1 teaspoon baking soda

* *To make the topping:* In a medium bowl, combine the flour, brown sugar, and cinnamon. Mix well. Add the butter. Mix until crumbly.

* *To make the muffins:* Preheat the oven to 400°F. Coat a 12-cup no-stick muffin tin with no-stick spray.

* In a medium bowl, combine the apples, dates, buttermilk, brown sugar, egg, applesauce, and oil. Mix well.

* In a large bowl, combine the flour, walnuts, baking powder, and baking soda. Mix well. Add the apple mixture. Stir until just blended. Pour into the prepared muffin tin, filling the cups three-fourths full. Sprinkle each muffin with some of the topping.

* Bake for 15 to 18 minutes, or until a toothpick inserted in the center of a muffin comes out clean.

Makes 12 muffins

Per muffin
Calories 160
Total fat 3.9 g.
Saturated fat 0.5 g.
Cholesterol 18 mg.
Sodium 177 mg.
Fiber 1.4 g.

Cost per serving

16¢

KITCHEN TIP

To freeze, place the cooled muffins on a tray. Put in the freezer for 1 hour, or until solid. Pack in a freezer-quality plastic bag. To use, thaw the number of muffins you need overnight in the refrigerator. Reheat at 400°F for 10 minutes, or until hot.

Yeast Breads, Muffins, and Quick Breads

CORNMEAL MUFFINS WITH BLUEBERRIES

Buttermilk and blueberries make these muffins tart and tasty—a nice balance to the brown sugar.

¾ cup low-fat buttermilk
½ cup packed brown sugar
¼ cup oil
1 egg
1 egg white
2 tablespoons unsweetened applesauce
1 cup all-purpose flour
½ cup yellow cornmeal
½ cup blueberries
1½ teaspoons baking powder
1 teaspoon baking soda
½ teaspoon ground cinnamon
¼ teaspoon salt

⊛ Preheat the oven to 400°F. Coat a 12-cup no-stick muffin tin with no-stick spray.

⊛ In a medium bowl, combine the buttermilk, brown sugar, oil, egg, egg white, and applesauce. Mix well.

⊛ In a large bowl, combine the flour, cornmeal, blueberries, baking powder, baking soda, cinnamon, and salt. Mix well. Add the buttermilk mixture. Stir until just blended. Pour into the prepared muffin tin, filling the cups three-fourths full.

⊛ Bake for 12 to 15 minutes, or until a toothpick inserted in the center of a muffin comes out clean.

Makes 12 muffins

Per muffin
Calories 150
Total fat 5.4 g.
Saturated fat 0.6 g.
Cholesterol 18 mg.
Sodium 242 mg.
Fiber 0.9 g.

Cost per serving

16¢

KITCHEN TIP

To freeze, place the cooled muffins on a tray. Put in the freezer for 1 hour, or until solid. Pack in a freezer-quality plastic bag. To use, thaw the number of muffins you need overnight in the refrigerator. Reheat at 400°F for 10 minutes, or until hot.

CHERRY STREUSEL COFFEE CAKE

This moist and fruity coffee cake will make a great contribution to your next potluck. If you have fresh cherries, by all means use them.

Makes 12 servings

Per serving
Calories 240
Total fat 6.6 g.
Saturated fat 0.7 g.
Cholesterol 18 mg.
Sodium 334 mg.
Fiber 1.4 g.

Cost per serving

15¢

Coffee Cake

¾ cup packed brown sugar
¾ cup low-fat buttermilk
½ cup unsweetened applesauce
⅓ cup prune puree
¼ cup oil
3 egg whites
1 egg
2 cups all-purpose flour
2 teaspoons baking powder
2 teaspoons baking soda
2 cups frozen cherries, thawed

Topping

¼ cup all-purpose flour
3 tablespoons packed brown sugar
½ teaspoon ground ginger
½ teaspoon ground cinnamon
1 tablespoon oil

※ *To make the coffee cake:* Preheat the oven to 350°F. Coat a 13" × 9" baking dish with no-stick spray.

※ In a medium bowl, combine the brown sugar, buttermilk, applesauce, prune puree, oil, egg whites, and egg. Mix well.

※ In a large bowl, combine the flour, baking powder, and baking soda. Mix well. Stir in the cherries. Add the buttermilk mixture. Stir until just blended.

※ Pour the batter into the prepared baking dish. Smooth the top with a spatula.

※ *To make the topping:* In a small bowl, combine the flour, brown sugar, ginger, cinnamon, and oil. Mix well. Sprinkle over the batter.

※ Bake for 40 to 45 minutes, or until a toothpick inserted in the center comes out clean. Cool before cutting.

KITCHEN TIP

To freeze, wrap the cooled coffee cake in freezer-quality plastic wrap, then in freezer-quality foil. To use, thaw overnight in the refrigerator. Remove the foil and plastic wrap; discard the plastic wrap. Rewrap in the foil. Reheat at 350° for 10 minutes, or until warm.

KITCHEN TIP

To freeze the cooled waffles, stack them, separated by small pieces of wax paper, and pack in a freezer-quality plastic bag. To use, remove from the freezer and reheat in a toaster oven.

CORN WAFFLES

A Midwestern favorite, these hearty waffles make a great supper or brunch entrée. Best of all, they can be frozen individually to pop into the toaster oven—saving you 15¢ apiece over store-bought toaster waffles. If you don't have a waffle iron, cook the batter on a griddle just like pancakes.

1½ cups low-fat buttermilk
¼ cup honey or sugar
2 eggs
1 tablespoon oil
1¼ cups all-purpose flour
½ cup yellow cornmeal
¼ cup frozen whole kernel corn, thawed
2 teaspoons baking powder
½ teaspoon baking soda
¼ teaspoon salt

❀ Coat a waffle iron with no-stick spray. Preheat the waffle iron.

❀ In a medium bowl, combine the buttermilk, honey or sugar, eggs, and oil. Mix well.

❀ In a large bowl, combine the flour, cornmeal, corn, baking powder, baking soda, and salt. Mix well. Add the buttermilk mixture. Stir until just blended.

❀ Spoon enough batter into the waffle iron to cover two-thirds of the surface. Bake for 5 minutes, or until the waffle is golden brown and cooked through. Repeat with the remaining batter. If serving immediately, keep the cooked waffles warm in the oven on a covered ovenproof plate.

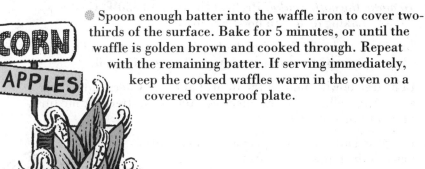

SPICED APPLE PANCAKES

Pancakes are everyone's favorite for weekend breakfasts. With the help of the freezer, you can enjoy this Sunday-morning special any day of the week.

1¾ cups all-purpose flour
1 cup low-fat buttermilk
1 small apple, grated
¼ cup low-fat cottage cheese
¼ cup oil
¼ cup packed brown sugar
1 egg
1 teaspoon baking soda
1 teaspoon ground cinnamon
¼ teaspoon allspice
¼ teaspoon salt

❋ In a blender, combine the flour, buttermilk, apples, cottage cheese, oil, brown sugar, egg, baking soda, cinnamon, allspice, and salt. Process until smooth. Set aside.

❋ Preheat a griddle or 10″ no-stick skillet over medium heat. Coat with no-stick spray.

❋ Spoon ¼ cup measures of the batter onto the griddle or skillet. Cook for 2 minutes, or until bubbles appear on the surface. Turn and cook for 2 minutes, or until golden brown and cooked through. Repeat to use all the batter. If serving immediately, keep the cooked pancakes warm in the oven on a covered oven-proof plate.

Makes 12 pancakes

Per pancake
Calories 148
Total fat 5.4 g.
Saturated fat 0.6 g.
Cholesterol 19 mg.
Sodium 197 mg.
Fiber 0.9 g.

Cost per serving

10¢

KITCHEN TIP

To freeze, place the cooled pancakes in a single layer on a tray. Freeze until solid. Stack, separated by pieces of wax paper. Place in a freezer-quality plastic bag. To use, remove from the freezer and reheat in a toaster oven.

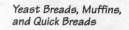

Yeast Breads, Muffins, and Quick Breads

CHAPTER 11

SAUCES AND CONDIMENTS

G ravies, sauces, chutneys, salsas, relishes—these are the flavor accessories that add so much style to simple grilled meats, roasted vegetables, and grain entrées.

It takes a bit of know-how to successfully freeze sauces and condiments. If you freeze a flour-thickened gravy, it will emerge separated and watery. Best bets for the freezer are cornstarch-thickened vegetable sauces such as our smooth and tangy Cheese Sauce.

Vegetable- and fruit-based sauces like Fall Fruit Ketchup and Chunky Pepper Pasta Sauce also freeze quite well. These recipes let you take advantage of lower prices when the produce harvest is bountiful.

And when you need to jazz up a meal fast, just turn to your freezer to whip up Marinara Sauce, Chili-Corn Chutney, or Sweet and Tangy Fruit Salsa.

Rounding out your selection of freezer-sauce essentials are meat and poultry marinades as well as low-fat dressings for salads.

251

Per ½ cup
Calories 44
Total fat 0.4 g.
Saturated fat 0 g.
Cholesterol 0 mg.
Sodium 30 mg.
Fiber 3.2 g.

Cost per serving

18¢

KITCHEN TIP

To freeze vegetable
pasta sauces, pack
them into small
freezer-quality
plastic containers.
To use, thaw
overnight in the
refrigerator. Place
in a saucepan,
cover, and cook,
stirring frequently,
over low heat, for
5 minutes, or until
just warm.

MARINARA SAUCE

This sauce features garlic galore, so adjust the amount according to personal preference. On an evening when the refrigerator is bare, this freezer sauce for pasta is a lifesaver.

1 cup frozen chopped onions
1 cup frozen chopped sweet red peppers
1 cup frozen broccoli florets
1 cup frozen sliced carrots
¼ cup frozen defatted Chicken Stock
     (page 61), thawed
1 can (28 ounces) reduced-sodium whole
     tomatoes, chopped (with juice)
1 can (6 ounces) reduced-sodium tomato paste
6 cloves garlic, crushed
2 teaspoons dried Italian herb seasoning
2 bay leaves

⊛ In a Dutch oven, combine the onions, peppers, broccoli, carrots, and stock. Cook, stirring, over medium-high heat for 5 minutes, or until soft but not browned. Add the tomatoes, tomato paste, garlic, Italian herb seasoning, and bay leaves. Bring to a boil.

⊛ Reduce the heat to medium. Cover and cook, stirring occasionally, for 30 minutes, or until thick. Remove and discard the bay leaves. Let cool.

⊛ Transfer to a blender or food processor. Process until smooth.

HERBED TOMATO SAUCE

What could be easier than a sauce that's not cooked? Make this summery sauce when tomatoes are abundant and ripe on the vine. Freeze for up to 6 months—the flavor actually improves in the freezer.

6 pounds tomatoes, coarsely chopped
1 teaspoon Italian herb seasoning
½ teaspoon ground black pepper
¼ teaspoon salt

※ In a blender or food processor, combine the tomatoes, Italian herb seasoning, black pepper, and salt. Process briefly to a coarse puree.

Makes 4 cups

Per ½ cup
Calories 72
Total fat 1.2 g.
Saturated fat 0.2 g.
Cholesterol 0 mg.
Sodium 96 mg.
Fiber 13.6 g.

Cost per serving
54¢

CHUNKY PEPPER PASTA SAUCE

With only a small time investment, you can enjoy this fine tomato pasta sauce long after the frost is on the pumpkin.

1 cup chopped onions
1 cup chopped sweet red peppers
1 tablespoon minced garlic
1 teaspoon olive oil
5 cups chopped tomatoes
¼ teaspoon dried oregano
2 tablespoons chopped parsley

※ In a Dutch oven, combine the onions, peppers, garlic, and oil. Cook, stirring, over medium-high heat for 5 minutes, or until soft but not browned. Add the tomatoes and oregano. Cook, stirring occasionally, for 10 minutes, or until thick. Add the parsley and stir well.

Makes 4 cups

Per ½ cup
Calories 42
Total fat 1 g.
Saturated fat 0.2 g.
Cholesterol 0 mg.
Sodium 14 mg.
Fiber 2 g.

Cost per serving
24¢

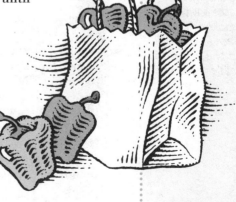

Sauces and Condiments

Per 1 tablespoon
Calories 22
Total fat 1.2 g.
Saturated fat 0.3 g.
Cholesterol 1 mg.
Sodium 26 mg.
Fiber 0.8 g.

Cost per serving

26¢

KITCHEN TIP

To freeze, place the pesto in ice-cube trays. Freeze the trays for 1 hour, or until solid. Transfer the cubes to a freezer-quality plastic bag. To use, thaw overnight in the refrigerator. Use 1 cube (1½ tablespoons) pesto for each 1 cup cooked pasta. Reserve 2 tablespoonfuls of the pasta cooking water to toss with the cooked drained pasta. That will help the pesto to better coat the pasta.

MINT-BASIL PESTO

Chefs recommend freezer pesto as a great way to preserve the pungent flavor of fresh herbs for months. Add this pesto by spoonfuls to salad dressings, sauces, and even bread dough. It's especially delicious on broiled chicken, lamb, and pork.

4   cups chopped fresh basil
1   cup chopped fresh mint
½   cup soft bread crumbs
¼   cup grated Parmesan cheese
2   tablespoons olive oil
2   teaspoons minced garlic
½   teaspoon ground black pepper

✸ In a blender or food processor, combine the basil, mint, bread crumbs, Parmesan, oil, garlic, and pepper. Process until smooth.

PLENTY OF PESTO

Pesto is one of the easiest toppings for pasta. This uncooked paste of chopped fresh herbs, garlic, and grated Parmesan keeps its quality in the freezer for up to six months.

Traditional basil pesto from the Italian city of Genoa includes plenty of high-fat olive oil, Parmesan cheese, and pine nuts. To lighten the fat load, substitute soft or dry bread crumbs for up to three-fourths of the oil. Add extra garlic and herbs to make up for any loss of flavor.

Prepare pesto during late summer, when herb crops are abundant. You can go beyond basil to include other herbs such as tarragon, cilantro, and parsley. Spinach leaves are good, too. Pack pesto in ice-cube trays and freeze until solid. Transfer the cubes to a freezer-quality plastic bag.

To season 1 pound cooked pasta, use about 10 cubes, which is just slightly less than 1 cup. To guarantee a creamy pesto that coats the pasta, reserve up to ½ cup of the pasta cooking liquid before draining. Toss the pasta with the pesto and enough cooking water to make the pesto cling to the pasta.

SPINACH AND PARMESAN PASTA TOPPING

It takes less time to make this pungent pasta topper than it does to cook the pasta. You can store a convenient supply of chopped parsley leaves in the freezer to have on hand for quick dishes like this.

2   cups packed frozen spinach
2   cups chopped fresh parsley
¼   cup grated Parmesan cheese
¼   cup frozen chopped onions, thawed
2   tablespoons minced garlic
2   tablespoons chopped walnuts or almonds
2   tablespoons olive oil
1   tablespoon dried oregano
⅛   teaspoon salt

❋ In a blender or food processor, combine the spinach, parsley, Parmesan, onions, garlic, walnuts or almonds, oil, oregano, and salt. Process until smooth.

Makes 1¼ cups

Per 1 tablespoon
Calories 33
Total fat 2.3 g.
Saturated fat 0.5 g.
Cholesterol 1 mg.
Sodium 56 mg.
Fiber 0.7 g.

Cost per serving

5¢

KITCHEN TIP

Use about 1½ tablespoons topping for each 1 cup cooked pasta. Reserve 2 tablespoonfuls of the pasta cooking water to toss with the cooked drained pasta. That will help the topping to better coat the pasta.

Makes 2 cups

Per 1/4 cup
Calories 20
Total fat 1.2 g.
Saturated fat 0.2 g.
Cholesterol 0 mg.
Sodium 74 mg.
Fiber 0.2 g.

Cost per serving

4¢

CHICKEN GRAVY

This simple gravy—made with frozen stock, frozen onions, and pantry staples—can be whipped up in 15 minutes. Its old-fashioned goodness graces roast chicken, turkey, or pork.

2   teaspoons olive oil
1/4 cup frozen chopped onions, minced
2   cloves garlic, minced
2   tablespoons all-purpose flour
2   cups frozen defatted Chicken Stock
        (page 61), thawed
1/4 teaspoon dried thyme
1/4 teaspoon ground black pepper
1/4 teaspoon salt

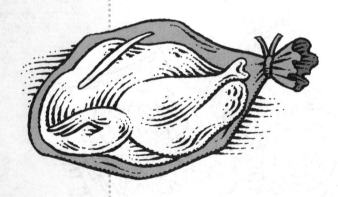

❋ Coat a 10″ no-stick skillet with no-stick spray and place over medium-high heat until hot. Add the oil, onions, and garlic. Cook, stirring, for 3 minutes, or until the onions are soft. Add the flour. Cook, stirring, for 2 minutes. Gradually stir in the stock.

❋ Bring to a boil. Reduce the heat to medium. Cook, stirring occasionally, for 10 minutes, or until the gravy thickens. Stir in the thyme, pepper, and salt.

CHEESE SAUCE

Most cheese sauces are too temperamental to freeze—but not this corn-starch-thickened version. Freeze in recipe-ready amounts to dress up everything from asparagus to ziti.

 2 tablespoons cornstarch
1½ cups skim milk
 ½ teaspoon dry mustard
1¼ cups shredded low-fat extra-sharp
     Cheddar cheese
 ½ cup low-fat ricotta cheese
 ¼ teaspoon salt
 ¼ teaspoon ground black pepper

✽ Place the cornstarch in a medium saucepan. Gradually add the milk and stir until smooth. Stir in the mustard. Bring to a boil, stirring frequently, over medium-high heat. Cook, stirring, for 3 minutes, or until the sauce thickens. Remove from the heat. Stir in the Cheddar until melted.

✽ In a blender or food processor, combine the ricotta, salt, and pepper. Process until smooth. Stir into the sauce.

Makes 3 cups

Per ¼ cup
Calories 60
Total fat 2.6 g.
Saturated fat 0.6 g.
Cholesterol 10 mg.
Sodium 148 mg.
Fiber 0 g.

Cost per serving

14¢

KITCHEN TIP

To freeze, place the cooled cooked sauce in small freezer-quality plastic containers or ice-cube trays. Freeze the trays for 1 hour, or until solid. Transfer the cubes to a freezer-quality plastic bag. To use, thaw overnight in the refrigerator. Puree briefly, if needed, to smooth. Place in a saucepan and cook over low heat until warm.

Sauces and Condiments

SPICY-HOT FRUIT SAUCE

KITCHEN TIP

To freeze hot peppers, just seed and mince (wear plastic gloves when handling). Freeze on a baking sheet to keep loose, then package in freezer-quality plastic bags. Hot peppers can be frozen for up to 5 months.

Don't be intimidated by the ingredient list. All the major ingredients are from the freezer, so the preparation goes quickly. This sweet-hot sauce adds excitement to lean grilled meats. If you have a small jalapeño pepper on hand, mince it and substitute it for the hot-pepper sauce.

½ cup frozen chopped onions
⅓ cup frozen chopped sweet red peppers
1 teaspoon olive oil
1½ cups frozen sliced rhubarb
1 cup frozen sliced peaches
1 cup frozen sliced strawberries
¾ cup honey or packed brown sugar
¾ cup cider vinegar
2 drops hot-pepper sauce
½ teaspoon curry powder
¼ teaspoon ground cinnamon
2 tablespoons minced fresh cilantro or parsley
⅛ teaspoon salt

✳ In a Dutch oven, combine the onions, red peppers, and oil. Cook, stirring, over medium heat for 5 minutes, or until the onions are soft but not browned. Add the rhubarb, peaches, strawberries, honey or brown sugar, vinegar, hot-pepper sauce, curry powder, and cinnamon. Bring to a boil. Cook for 10 minutes, or until the fruit is very soft. Stir in the cilantro or parsley and salt.

RASPBERRY DESSERT SAUCE

Try this French-style sauce with angel food cake, lemon sorbet, or low-fat chocolate frozen yogurt. It's an elegant addition to any meal for less than $1 per serving and 5 minutes of kitchen time.

4 cups frozen raspberries, thawed
⅓ cup packed brown sugar
2 tablespoons lemon juice
½ teaspoon ground nutmeg

❋ In a blender or food processor, combine the raspberries, brown sugar, lemon juice, and nutmeg. Process until smooth.

❋ Place a fine sieve over a medium bowl. With the back of a large spoon or a rubber spatula, press the puree through the sieve. Discard the seeds.

Makes 2 cups

Per ¼ cup
Calories 66
Total fat 0.4 g.
Saturated fat 0 g.
Cholesterol 0 mg.
Sodium 4 mg.
Fiber 2.5 g.

Cost per serving
96¢

FRUITS ARE TOPS

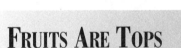

When fruits are plentiful and sweet, capture their essence with these chunky sauces for hot cereals, pancakes, waffles, English muffins, or toast. Each yields about 1 cup and freezes for up to 2 months.

❋ *Apple.* In a medium saucepan, combine 2 cups chopped apples, 2 tablespoons honey or brown sugar, and 1 teaspoon lemon juice. Cover and cook, stirring occasionally, over low heat for 15 minutes, or until very soft. Mash well.

❋ *Berry.* In a blender or food processor, process 1 cup blueberries, raspberries, or strawberries with 1 tablespoon brown sugar and 1 teaspoon lemon juice until smooth.

❋ *Pineapple.* In a blender or food processor, process 1 cup chopped pineapple with 2 tablespoons brown sugar and ½ teaspoon vanilla until smooth.

Sweet Orange Syrup

Have a container of this sweet and tangy orange syrup on hand in the freezer to pour over waffles, pancakes, or low-fat frozen yogurt.

1   cup orange marmalade
¼   cup orange juice
3   tablespoons packed brown sugar
1   tablespoon lemon juice

❋ In a small saucepan, combine the marmalade, orange juice, brown sugar, and lemon juice. Cook, stirring, over medium-high heat for 3 minutes, or until the marmalade melts.

❋ Place a fine sieve over a medium bowl. With the back of a large spoon or a rubber spatula, press the mixture through the sieve. Discard any orange rind from the marmalade.

Kitchen Tip

To freeze the cooled syrup, pack into small freezer-quality plastic containers. To use, thaw overnight in the refrigerator. Serve at room temperature or microwave until warm.

Mustardy Marinade

Frozen vegetables and spices combine in this mustard-based marinade for fish. It can transform an ordinary main dish into an extraordinary entrée for only 5¢ a serving.

⅓   cup packed brown sugar
¼   cup frozen chopped onions
¼   cup frozen sliced peaches
2   tablespoons Dijon mustard
2   tablespoons reduced-sodium soy sauce
1   teaspoon minced garlic
¼   teaspoon ground red pepper

❋ In a blender or food processor, combine the brown sugar, onions, peaches, mustard, soy sauce, garlic, and pepper. Process until smooth.

Jerk Marinade

Use this spicy Jamaican seasoning paste to marinate 2 pounds of boneless chicken breasts or firm fish fillets overnight in the refrigerator before broiling or grilling.

⅓ cup frozen chopped sweet red peppers, minced
¼ cup frozen chopped onions, minced
1 small jalapeño pepper, minced
    (wear plastic gloves when handling)
1 tablespoon minced garlic
1 tablespoon minced fresh ginger
1 tablespoon reduced-sodium soy sauce
2 teaspoons oil
½ teaspoon dried thyme
¼ teaspoon allspice
¼ teaspoon ground black pepper

❃ In a medium bowl, combine the red peppers, onions, jalapeño peppers, garlic, ginger, soy sauce, oil, thyme, allspice, and black pepper. Stir well.

Makes 1 cup

Per 1 tablespoon
Calories 10
Total fat 0.6 g.
Saturated fat 0 g.
Cholesterol 0 mg.
Sodium 38 mg.
Fiber 0.1 g.

Cost per serving
4¢

Fruity Barbecue Sauce

Frozen plums or peaches liven up this Southern barbecue sauce. It's mild and sweet, perfect for basting or for serving over pork or chicken. And at about $1 less than supermarket sauces, it's a real bargain.

3 cups frozen sliced plums or peaches
1 cup frozen chopped onions
¼ cup orange juice
2 tablespoons honey
2 teaspoons reduced-sodium soy sauce
1 teaspoon minced garlic
½ teaspoon ground cinnamon

❃ In a medium saucepan, combine the plums or peaches, onions, orange juice, and honey. Bring to a boil over medium-high heat. Reduce the heat to medium. Cook, stirring occasionally, for 10 minutes, or until very soft. Let cool. Transfer to a blender or food processor. Add the soy sauce, garlic, and cinnamon. Process until smooth.

Makes 1½ cups

Per 2 tablespoons
Calories 42
Total fat 0.3 g.
Saturated fat 0 g.
Cholesterol 0 mg.
Sodium 36 mg.
Fiber 1.1 g.

Cost per serving
14¢

Sauces and
Condiments

Per 2 tablespoons
Calories 27
Total fat 1.6 g.
Saturated fat 0.2 g.
Cholesterol 0 mg.
Sodium 79 mg.
Fiber 0.2 g.

Cost per serving

6¢

KITCHEN TIP

To freeze, place the dressing in ice-cube trays. Freeze for 1 hour, or until solid. Transfer the cubes to a freezer-quality plastic bag. To use, thaw overnight in the refrigerator.

MUSTARD-GARLIC DRESSING

This salad dressing can also be used as a basting sauce or a marinade for grilled pork or chicken.

¼ cup chopped onions
¼ cup balsamic vinegar
1 tablespoon olive oil
1 tablespoon frozen apple juice concentrate
2 tablespoons Dijon mustard
2 tablespoons minced garlic
1 teaspoon lemon juice
1 tablespoon minced fresh parsley

✳ In a blender or food processor, combine the onions, vinegar, oil, apple juice concentrate, mustard, garlic, and lemon juice. Process until smooth. Stir in the parsley.

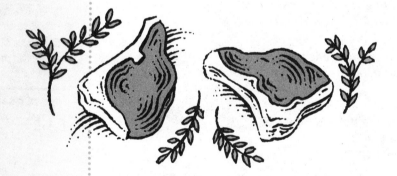

CREAMY LIME SALAD DRESSING

Use this creamy sauce to dress up fruit salads or mixed greens. It costs about $1.25 less than most low-fat supermarket dressings. Keep on hand in the freezer in meal-size portions.

1 cup nonfat plain yogurt
½ cup nonfat cottage cheese
2 tablespoons lime juice
2 tablespoons packed brown sugar
1 teaspoon oil

❋ In a blender or food processor, combine the yogurt, cottage cheese, lime juice, brown sugar, and oil. Process until smooth.

CREAMY HERB SALAD DRESSING

Serve this salad dressing as a dip for fresh vegetables. Use basil, parsley, tarragon, thyme, or whatever fresh herbs are available.

1 cup nonfat cottage cheese
¼ cup nonfat plain yogurt
2 tablespoons lemon juice
2 tablespoons crumbled blue cheese
¼ teaspoon curry powder
¼ teaspoon ground black pepper
1–2 tablespoons minced fresh herbs

❋ In a blender or food processor, combine the cottage cheese, yogurt, lemon juice, blue cheese, curry powder, and pepper. Process until smooth. Stir in the herbs.

Makes 1½ cups

Per 2 tablespoons
Calories 30
Total fat 0.4 g.
Saturated fat 0.1 g.
Cholesterol 1 mg.
Sodium 43 mg.
Fiber 0 g.

Cost per serving

11¢

KITCHEN TIP

To freeze these dressings, place in ice-cube trays. Freeze for 1 hour, or until solid. Transfer the cubes to a freezer-quality plastic bag. To use, thaw overnight in the refrigerator. Process briefly in a blender or food processor until smooth.

Makes 1⅓ cups

Per 2 tablespoons
Calories 24
Total fat 0.5 g.
Saturated fat 0.3 g.
Cholesterol 3 mg.
Sodium 92 mg.
Fiber 0 g.

Cost per serving

15¢

Sauces and Condiments

Per 2 tablespoons
Calories 37
Total fat 1.4 g.
Saturated fat 0.5 g.
Cholesterol 2 mg.
Sodium 163 mg.
Fiber 0 g.

Cost per serving

16¢

KITCHEN TIP

To freeze, place the dressing in ice-cube trays. Freeze the trays for 1 hour, or until solid. Transfer the cubes to a freezer-quality plastic bag. To use, thaw overnight in the refrigerator. Process briefly in a blender or food processor until smooth.

PARMESAN-PEPPER DRESSING

Especially delicious over bitter lettuce such as endive or over shredded cabbage, this coleslaw dressing saves you 50¢ over the same amount of store-bought dressing.

½ cup low-fat mayonnaise
½ cup low-fat cottage cheese
⅓ cup low-fat buttermilk
⅓ cup grated Parmesan cheese
2 tablespoons lemon juice
2 teaspoons sugar
1 clove garlic, minced
½ teaspoon ground black pepper

✻ In a blender or food processor, combine the mayonnaise, cottage cheese, buttermilk, Parmesan, lemon juice, sugar, garlic, and pepper. Process until smooth.

FREEZER HELPERS

Chef Michael Roberts, author of *Fresh from the Freezer*, recommends freezing a variety of condiments as helpers to make dishes sparkle when there's only time for the most basic cooking. Roberts stocks pestos, chutneys, pasta fillings, stuffings, stocks, tomato sauce, and chopped onions. Ingredients that he might otherwise omit from a recipe because they add 15 minutes to dinner preparation can be thawed in seconds in the microwave.

"A stockpile of these freezer helpers allows me to toss together a seemingly complicated meal with very little effort," Roberts says. "Ordinary fare becomes the stuff of epicures."

SWEET AND TANGY FRUIT SALSA

This unusual fruit salsa is an easy topping for broiled fish, pork, or chicken.

1 cup frozen chopped onions
3 cloves garlic, chopped
1 teaspoon olive oil
1 cup frozen blueberries
1 cup frozen cherries, chopped
¼ cup orange juice
2 tablespoons sugar
1 teaspoon curry powder
1 teaspoon ground red pepper
¼ teaspoon salt
2 tablespoons chopped fresh cilantro or parsley

✳ In a medium saucepan, combine the onions, garlic, and oil. Cook, stirring, over medium heat for 5 minutes, or until soft but not browned. Add the blueberries, cherries, orange juice, sugar, curry powder, red pepper, and salt. Bring to a boil. Cook, stirring occasionally, for 5 minutes, or until the liquid begins to thicken. Stir in the cilantro or parsley.

Makes 2 cups

Per ¼ cup
Calories 70
Total fat 0.9 g.
Saturated fat 0.1 g.
Cholesterol 0 mg.
Sodium 71 mg.
Fiber 1.5 g.

Cost per serving

29¢

KITCHEN TIP

Get full value from any bunch of herbs you purchase. After using the amount you need for a recipe, chop and freeze the remaining leaves in a small freezer-quality plastic bag. Cilantro, parsley, and other herbs will remain green and flavorful. Use the herbs while still frozen.

Per ¼ cup
Calories 66
Total fat 4.1 g.
Saturated fat 0.6 g.
Cholesterol 0 mg.
Sodium 33 mg.
Fiber 0.7 g.

Cost per serving

23¢

KITCHEN TIP

To freeze hot peppers, just seed and mince (wear plastic gloves when handling). Freeze on a baking sheet to keep loose, then package in freezer-quality plastic bags. Hot peppers can be frozen for up to 5 months.

SWEET AND SPICY STRAWBERRY SALSA

The sweetness of balsamic vinegar compensates for less oil in this refreshing berry condiment.

2   cups frozen sliced strawberries
½   cup balsamic vinegar
1   small jalapeño pepper, minced
     (wear plastic gloves when handling)
3   tablespoons olive oil
2   tablespoons lemon juice
2   tablespoons minced fresh mint
2   teaspoons sugar
⅛   teaspoon salt
⅛   teaspoon ground black pepper

✳ In a medium bowl, combine the strawberries, vinegar, jalapeño peppers, oil, lemon juice, mint, sugar, salt, and black pepper. Mix well. Set aside for 15 minutes, or until the strawberries are thawed. Stir to mix.

TOMATILLO SALSA

Tomatillos—small, husk-covered green fruits that are related to toma-toes—freeze best if cooked and slightly reduced by making them into a sauce or salsa. This Texas-style topping is traditionally served over enchiladas or alongside barbecued meats. If you don't have tomatillos, substitute green tomatoes.

4 cups chopped tomatillos
2 cups chopped onions
¼ cup chopped sweet red peppers
1 small jalapeño pepper, minced
      (wear plastic gloves when handling)
1 teaspoon olive oil
¼ cup lemon juice
¼ cup chopped fresh parsley
½ teaspoon salt

✳ In a large saucepan, combine the tomatillos, onions, red peppers, jalapeño peppers, and oil. Cook, stirring. Bring to a boil over medium-high heat. Cook, stirring, for 10 minutes, or until reduced by half. Stir in the lemon juice, parsley, and salt.

ROASTED PEPPERS

Roasted peppers—both hot and sweet—are one of the most versatile condiments that you can have in your freezer. Keep them on hand to top scrambled eggs, grilled shrimp, poultry, pork, or beef. Or toss chopped roasted peppers into cooked grains or beans.

To roast peppers, halve them lengthwise and remove the seeds. Place, cut side down, on a broiler pan. Broil 4" from the heat until blistered. Transfer to a paper bag, close, and set aside to steam for 10 or 15 minutes. This step makes the thin, tough skin easier to remove. Peel off the skin and discard. Freeze in freezer-quality plastic bags or freezer-quality containers for up to six months. Thaw overnight in the refrigerator before using.

Makes 3 cups

Per ¼ cup
Calories 32
Total fat 0.8 g.
Saturated fat 0 g.
Cholesterol 0 mg.
Sodium 94 mg.
Fiber 1.6 g.

Cost per serving
26¢

KITCHEN TIP

To freeze, pack the cooled cooked salsa in small freezer-quality plastic containers. To use, thaw overnight in the refrigerator.

Sauces and Condiments

CHILI-CORN CHUTNEY

Makes 2 cups

Per ¼ cup
Calories 57
Total fat 0.7 g.
Saturated fat 0.1 g.
Cholesterol 0 mg.
Sodium 7 mg.
Fiber 1.2 g.

Cost per serving

17¢

Frozen corn makes a delicious chutney when combined with peppers, onions, spices, and ginger. This colorful concoction is an outstanding complement to roast turkey, chicken, or mixed root vegetables.

1½ cups frozen whole kernel corn
½   cup frozen chopped onions
¼   cup frozen chopped sweet red peppers
2   cloves garlic, minced
1   teaspoon olive oil
3   tablespoons cider vinegar
2   tablespoons honey
1   tablespoon chopped fresh ginger
1   teaspoon chili powder
¼   teaspoon curry powder
1   teaspoon cornstarch
2   tablespoons apple juice
2   tablespoons minced fresh parsley

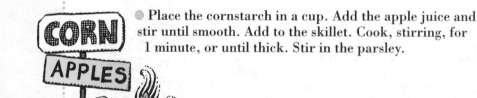

※ In a 10″ no-stick skillet, combine the corn, onions, peppers, garlic, and oil. Cook, stirring, over medium-high heat for 5 minutes, or until the onions are soft but not browned. Add the vinegar, honey, ginger, chili powder, and curry powder. Cook, stirring occasionally, for 5 minutes, or until the corn is tender.

※ Place the cornstarch in a cup. Add the apple juice and stir until smooth. Add to the skillet. Cook, stirring, for 1 minute, or until thick. Stir in the parsley.

Orange-Onion Relish

A zesty addition to grilled meats or burgers, this relish also makes an economical and low-fat sandwich spread.

1 cup frozen chopped onions, minced
½ cup frozen chopped sweet red peppers, minced
¼ cup frozen orange juice concentrate
¼ cup chopped scallions
2 tablespoons olive oil
1 tablespoon honey
1 teaspoon Dijon mustard
½ teaspoon ground black pepper

❋ In a medium bowl, combine the onions, red peppers, orange juice concentrate, scallions, oil, honey, mustard, and black pepper. Mix well. Let stand at room temperature for 30 minutes, stirring occasionally.

Makes 1½ cups

Per 2 tablespoons
Calories 37
Total fat 2.3 g.
Saturated fat 0.3 g.
Cholesterol 0 mg.
Sodium 13 mg.
Fiber 0.6 g.

Cost per serving
10¢

Beet Relish

Frozen beets are already peeled, eliminating the mess involved with preparing fresh beets. Roasting the beets turns them into sweet nuggets.

1 pound frozen beets
¼ cup frozen chopped onions, minced
¼ cup raisins
1 tablespoon cider vinegar
1 tablespoon olive oil
1 tablespoon chopped fresh parsley
¼ teaspoon ground black pepper

❋ Preheat the oven to 350°F. Wrap the beets in foil. Roast for 1 hour, or until very tender. Let cool, then chop finely.

❋ Transfer to a medium bowl. Add the onions, raisins, vinegar, oil, parsley, and pepper. Mix well.

Makes 2 cups

Per 2 tablespoons
Calories 28
Total fat 0.9 g.
Saturated fat 0.2 g.
Cholesterol 0 mg.
Sodium 23 mg.
Fiber 0.7 g.

Cost per serving
20¢

Per 2 tablespoons
Calories 23
Total fat 1.2 g.
Saturated fat 0.2 g.
Cholesterol 0 mg.
Sodium 29 mg.
Fiber 0.8 g.

Cost per serving

9¢

KITCHEN TIP

To freeze, pack the cooled cooked relish in small freezer-quality plastic containers. To use, thaw overnight in the refrigerator.

ROAST VEGETABLE RELISH

Roasting vegetables is a hands-off procedure that yields extraordinary sweetness and depth of flavor to this dish.

2	cups halved cherry tomatoes
2	cups chopped onions
2	cups chopped sweet red peppers
1	cup sliced carrots
2	tablespoons olive oil
1	tablespoon minced garlic
½	teaspoon Italian herb seasoning
1	tablespoon balsamic vinegar
¼	teaspoon salt

✤ Preheat the oven to 350°F.

✤ Place the tomatoes, onions, peppers, and carrots on a large piece of foil. Sprinkle with the oil, garlic, and Italian herb seasoning. Fold the foil over the vegetables to form a packet. Roast for 1 hour, or until very tender.

✤ Let cool, then chop finely. Place in a large bowl. Add the vinegar and salt. Toss to combine.

FALL FRUIT KETCHUP

Apples and tomatoes permeated with sweet spices make an old-fashioned condiment that tastes wonderful with barbecued meats, especially pork. The texture is slightly rougher and the color more subtle than mass-produced supermarket ketchup.

 6 medium tomatoes, chopped
 2 cups chopped onions
 2 large tart apples, peeled and chopped
 2 pears, peeled and chopped
 1 cup chopped celery
¾ cup apple cider vinegar
⅓ cup packed brown sugar
½ teaspoon ground allspice
½ teaspoon ground cinnamon
¼ teaspoon ground cloves

✸ In a Dutch oven, combine the tomatoes, onions, apples, pears, celery, vinegar, brown sugar, allspice, cinnamon, and cloves. Cook, stirring occasionally, over medium heat for 1 hour, or until thick. Let cool. Transfer to a blender or food processor. Process until smooth.

Makes 7 cups

Per 2 tablespoons
Calories 18
Total fat 0.1 g.
Saturated fat 0 g.
Cholesterol 0 mg.
Sodium 5 mg.
Fiber 0.6 g.

Cost per serving

6¢

KITCHEN TIP

To freeze, place the cooled cooked ketchup in ice-cube trays. Freeze the trays for 1 hour, or until solid. Transfer the cubes to a freezer-quality plastic bag. To use, thaw overnight in the refrigerator. Puree briefly, if needed, to smooth.

CHAPTER 12

CAKES, COOKIES, AND OTHER DESSERTS

Homemade desserts finish off a meal with a flourish. Freeze home-baked goodies when you have spare time. It'll mean time saved when you are rushed, money saved over store-bought, and smiles around the dining table.

Cakes and sweet breads—which have a relatively low moisture content— freeze without noticeable change in taste or texture. Most fruit desserts, such as cobblers, crisps, and pies, can be frozen baked or unbaked. Since it takes little more time to assemble three fruit desserts instead of one, you'd be wise to stock the freezer.

Cookies can be frozen for up to four months, and they thaw in just 10 minutes. You can even reheat them briefly in a 350°F oven so they'll taste freshly baked. You can also freeze unbaked cookie dough in a log shape, which is $1.50 to $2.50 cheaper per package than commercial refrigerated cookie dough. It will keep for about 4 months.

Per serving
Calories 124
Total fat 0.4 g.
Saturated fat 0.2 g.
Cholesterol 0 mg.
Sodium 47 mg.
Fiber 1.2 g.

Cost per serving

12¢

KITCHEN TIP

To freeze, wrap the cooled cake in freezer-quality plastic wrap, then in freezer-quality foil. To use, unwrap and cover loosely with paper towels before thawing overnight in the refrigerator.

CHOCOLATE ANGEL FOOD CAKE

Angel food cake rises to new heights with this chocolate-rich recipe. Serve it with fresh berries or low-fat vanilla yogurt.

1	cup all-purpose flour
1	cup sugar
⅓	cup cocoa powder
½	teaspoon ground cinnamon
10	egg whites, at room temperature
1¼	teaspoons cream of tartar
1½	teaspoons vanilla

❋ Preheat the oven to 350°F. Sift the flour and ½ cup of the sugar into a medium bowl. Add the cocoa and cinnamon. Mix well.

❋ Place the egg whites in a large bowl. Beat with an electric mixer until foamy. Add the cream of tartar and beat until soft peaks form. Gradually beat in the remaining ½ cup sugar; continue beating until stiff peaks form.

❋ Fold in the flour mixture ½ cup at a time. Pour into a 10" tube pan, spreading evenly and deflating any large air pockets with a knife.

❋ Bake for 40 minutes, or until a toothpick inserted in the center comes out clean. Cool upside down for 40 minutes before removing the cake from the pan.

Wrap It Up!

How you wrap your baked goods
for the freezer determines how
fresh they taste after thawing, says
Elinor Klivans, author of *Bake and Freeze
Chocolate Desserts*. Cool each item completely at room temperature, but never refrigerate because that will dry the
item.

Muffins, cupcakes, and other delicate items can be flash-
frozen before wrapping to prevent crumbling during freezer
storage. Place the items on a tray and set in the freezer for
several hours, or until frozen solid.

Wrap cooled or flash-frozen baked goods tightly in freezer-
quality plastic wrap, then with freezer-quality foil. Label and
date the outside of the foil. Cupcakes and cookies can be
wrapped in plastic, then stored in airtight tins or containers
instead of being wrapped in foil. Cheesecake freezes best as
individually wrapped slices.

Remember to store baked goods with other fragile items.
Don't set a 20-pound turkey on top of chocolate cupcakes!

Per serving
Calories 160
Total fat 3.5 g.
Saturated fat 1.8 g.
Cholesterol 58 mg.
Sodium 281 mg.
Fiber 1.3 g.

Cost per serving

14¢

KITCHEN TIP

To freeze, wrap the cooled cake in freezer-quality plastic wrap, then in freezer-quality foil. To use, thaw overnight in the refrigerator.

CHOCOLATE SPICE CAKE

With a chocolate cake in the freezer, a birthday celebration can be a carefree occasion for the cook as well as the guests. Frost the cake after thawing. To make a layer cake, double the recipe and bake in 2 round pans (8" diameter).

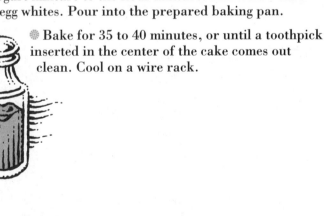

1   cup all-purpose flour
½   cup sugar
3   tablespoons cocoa powder
1   teaspoon ground cinnamon
1   teaspoon baking soda
¼   teaspoon salt
2   egg yolks
1   cup nonfat plain yogurt
2   tablespoons reduced-calorie butter, melted
1   teaspoon vanilla
2   egg whites

❋ Preheat the oven to 350°F. Coat an 8" × 8" freezer-proof baking pan with no-stick spray.

❋ In a large bowl, combine the flour, sugar, cocoa, cinnamon, baking soda, and salt. Mix well.

❋ In a medium bowl, combine the egg yolks, yogurt, butter, and vanilla. Mix well.

❋ Place the egg whites in a small bowl. Beat with an electric mixer until stiff peaks form.

❋ Add the yogurt mixture to the flour mixture. Mix well. Fold in the egg whites. Pour into the prepared baking pan.

❋ Bake for 35 to 40 minutes, or until a toothpick inserted in the center of the cake comes out clean. Cool on a wire rack.

PINEAPPLE UPSIDE-DOWN CAKE

Oatmeal adds a pleasantly chewy texture to this old-fashioned favorite. The cake is great served with low-fat vanilla frozen yogurt.

  1 can (8 ounces) pineapple rings packed in juice
  2 teaspoons reduced-calorie butter
1¼ cups packed brown sugar
1¼ cups all-purpose flour
  ⅓ cup rolled oats
  1 teaspoon baking powder
  ½ teaspoon baking soda
  ½ teaspoon ground ginger
  ½ teaspoon ground cinnamon
  ½ cup low-fat buttermilk
  1 egg
  1 teaspoon vanilla

* Preheat the oven to 350°F.

* Drain the pineapple, reserving the juice. Arrange the pineapple rings in a single layer in a 9" round cake pan.

* In a small saucepan, combine the butter, ½ cup of the brown sugar, and ¼ cup of the reserved pineapple juice. Cook, stirring, over medium heat for 3 minutes, or until bubbling. Pour over the pineapple rings.

* In a 10" no-stick skillet, combine the flour and oats. Cook, stirring, over medium-high heat for 3 minutes, or until light brown and fragrant. Remove from the heat and transfer to a medium bowl. Stir in the baking powder, baking soda, ginger, cinnamon, and the remaining ¾ cup brown sugar. Mix well.

* In a small bowl, combine the buttermilk, egg, and vanilla. Mix well. Add to the flour mixture. Mix well. Pour over the pineapple rings.

* Bake for 35 minutes, or until a toothpick inserted in the center of the cake comes out clean. Let cool for 5 minutes. Using a knife, loosen the cake from the sides of the pan. Invert onto a plate.

Makes 8 servings

Per serving
Calories 251
Total fat 1.7 g.
Saturated fat 0.7 g.
Cholesterol 29 mg.
Sodium 184 mg.
Fiber 1.3 g.

Cost per serving

24¢

KITCHEN TIP

To freeze, place the cooled cake on a freezer-proof plate and wrap in freezer-quality plastic wrap, then in freezer-quality foil. To use, thaw overnight in the refrigerator.

Cakes, Cookies, and Other Desserts

Per serving
Calories 240
Total fat 3.6 g.
Saturated fat 1.6 g.
Cholesterol 78 mg.
Sodium 81 mg.
Fiber 1.3 g.

Cost per serving

52¢

CHERRY PUDDING CAKE

In France, soft cakes like this are called clafouti, and they're often made with tart cherries harvested during July and August. Instead of traditional heavy cream, we substituted yogurt to lighten the batter.

2   cups frozen dark sweet cherries, thawed
1   cup nonfat plain yogurt
½   cup part-skim ricotta cheese
¼   cup all-purpose flour
¼   cup sugar
2   eggs
¼   teaspoon ground nutmeg
¼   cup raisins
2   tablespoons packed brown sugar

✳ Preheat the oven to 400°F. Coat a 12″ quiche dish with no-stick spray. Place the cherries in the dish.

✳ In a blender or food processor, combine the yogurt, ricotta, flour, sugar, eggs, and nutmeg. Process until smooth. Pour over the cherries. Sprinkle with the raisins.

✳ Bake for 12 minutes, or until the cake begins to set. Sprinkle with the brown sugar. Bake for 8 minutes, or until the cake is firm and golden brown.

TO FREEZE OR NOT TO FREEZE

Elinor Klivans, author of *Bake and Freeze Chocolate Desserts*, gives the following guidelines for which desserts freeze well and which don't.

- Densely textured desserts, like cookies or unfrosted cakes, freeze longer and better than mousses, puddings, and other very moist desserts. High moisture turns to ice crystals, which break up the dessert's smooth texture. Freeze cakes in layers and cookies in small packages.

- Desserts that are good keepers in the refrigerator—such as apple cobbler—are also good choices for freezing.

- Sauces thickened with cornstarch or tapioca freeze well. Flour-based sauces separate and become grainy in the freezer.

- Pie and tart crusts destined to have an uncooked filling or a filling that's cooked separately freeze better unbaked. No need to thaw before baking. Just add 5 to 10 minutes to the baking time.

- Filled pies, such as apple, raspberry, cherry, peach, blackberry, cranberry, and blueberry, freeze better unbaked. No need to thaw before baking. Just add 10 to 15 minutes to the baking time.

- Wrap all desserts in freezer-quality plastic wrap and freezer-quality foil to prevent moisture loss during freezing.

Cakes, Cookies, and Other Desserts

WARM HEARTS WITH FROZEN TREATS

Many of the freezer cookies, muffins, and
quick breads in this book make delectable gifts.
Here are some ways to pack them for holidays, housewarm-
ings, and other special occasions.

* *Cookies*. Pack cookies in a rigid container to prevent crumbling. Elinor Klivans, author of *Bake and Freeze Chocolate Desserts*, recommends wrapping two to four cookies in
  plastic, then arranging the packages in a decorative tin or
  container. This makes it easier to thaw a small amount at a
  time and cuts down on waste. On a small note card, write
  directions to crisp the cookies in a 350°F oven for 10 minutes.

* *Fruit cobblers and crisps*. Cobblers and crisps can be
  frozen right in a decorative baking pan, then wrapped in
  foil. Tie a pretty dish towel around the pan and attach a
  card with reheating directions.

* *Meringues*. Freeze baked meringues (page 288) in a decorative tin. Include a container of sweetened home-frozen
  berries for topping.

* *Muffins*. Flash-freeze, then store in a single layer in
  freezer-quality plastic bags. Expel all excess air. Include
  reheating directions.

* *Pies*. Freeze fruit pies unbaked. Add an additional 2 tablespoons bread crumbs or uncooked instant tapioca to the
  filling to absorb the fruit's moisture. Stir the bread crumbs
  or tapioca into the filling before pouring into the prepared
  pie crust. Include cooking directions; it's best to bake the
  pies while still frozen, adding 10 minutes to the total
  cooking time.

Gingerbread with Honey-Lemon Sauce

Gingerbread freezes beautifully and can be reheated in a flash for company or family desserts. The lemon sauce can be made in a moment just before serving.

Gingerbread

- 1 cup molasses
- 1 cup low-fat buttermilk
- ½ cup packed brown sugar
- ¼ cup oil
- ⅓ cup unsweetened applesauce
- 2 eggs
- 2¼ cups all-purpose flour
- 1 tablespoon ground ginger
- 2 teaspoons baking soda
- ½ teaspoon ground cloves

Sauce

- 1 cup nonfat plain yogurt
- 2 tablespoons honey
- 1 teaspoon grated lemon rind

❋ **To make the gingerbread:** Preheat the oven to 350°F. Coat a 10″ tube pan with no-stick spray. Sprinkle with flour and shake out any excess.

❋ In a medium bowl, combine the molasses, buttermilk, brown sugar, oil, applesauce, and eggs. Mix well.

❋ In a large bowl, sift together the flour, ginger, baking soda, and cloves. Add the molasses mixture. Stir just enough to blend. Pour into the prepared tube pan.

❋ Bake for 35 minutes, or until a toothpick inserted in the center comes out clean. Let cool before removing from the pan.

❋ **To make the sauce:** In a small bowl, combine the yogurt, honey, and lemon rind. Mix well. To serve, slice the gingerbread and place on dessert plates. Spoon the sauce over the cake or next to it.

Makes 12 servings

Per serving
Calories 278
Total fat 5.9 g.
Saturated fat 0.7 g.
Cholesterol 37 mg.
Sodium 272 mg.
Fiber 1 g.

Cost per serving

27¢

KITCHEN TIP

To freeze, wrap the cooled cake in freezer-quality plastic wrap, then in freezer-quality foil. To use, thaw overnight in the refrigerator. Reheat at 350°F for 10 minutes, or until warm.

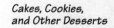

Per cupcake
Calories 201
Total fat 8.2 g.
Saturated fat 2.2 g.
Cholesterol 36 mg.
Sodium 110 mg.
Fiber 2.4 g.

Cost per serving

23¢

KITCHEN TIP

To freeze, place the cooled cupcakes on a tray. Put in the freezer for 1 hour, or until solid. Transfer to freezer-quality plastic bags. Press out the air and seal well. To use, thaw overnight in the refrigerator.

Per cookie
Calories 49
Total fat 0.8 g.
Saturated fat 0.6 g.
Cholesterol 2 mg.
Sodium 36 mg.
Fiber 0.3 g.

Makes 36 cookies

Cost per serving

3¢

CINNAMON-CHOCOLATE CUPCAKES

Mashed bananas reduce the fat and lighten these chocolate cupcakes. Choose any frosting from "The Finishing Touch" for a special topping.

- 1 cup all-purpose flour
- ¾ cup sugar
- ½ cup cocoa powder
- 1½ teaspoons baking powder
- 1 teaspoon ground cinnamon
- ¼ teaspoon baking soda
- ½ cup semisweet chocolate chips
- ½ cup low-fat buttermilk
- ¼ cup oil
- 2 small ripe bananas
- 2 eggs
- 1 teaspoon vanilla

❈ Preheat the oven to 350°F. Coat a 12-cup muffin tin with no-stick spray.

❈ In a large bowl, combine the flour, sugar, cocoa, baking powder, cinnamon, and baking soda. Mix well. Stir in the chocolate chips.

❈ In a blender or food processor, combine the buttermilk, oil, bananas, eggs, and vanilla. Process until smooth. Stir into the flour mixture. Spoon into the muffin cups, filling the cups three-quarters full.

❈ Bake for 35 minutes, or until a toothpick inserted in the center of 1 cupcake comes out clean. Cool on a wire rack.

GINGER COOKIES

These cookies are tasty for school lunches or holiday gifts. Top them with your choice of glaze from "The Finishing Touch."

- 2 cups all-purpose flour
- ¾ teaspoon baking soda
- ½ teaspoon ground ginger
- ½ teaspoon ground cinnamon
- ¼ teaspoon ground cloves
- ½ cup packed brown sugar
- ¼ cup molasses
- ¼ cup reduced-calorie butter, at room temperature

- In a large bowl, combine the flour, baking soda, ginger, cinnamon, and cloves. Mix well.

- In a blender or food processor, combine the brown sugar, molasses, and butter. Process until smooth. Stir into the flour mixture.

- Form the dough into a log. Wrap in plastic wrap and refrigerate for 30 minutes.

- Preheat the oven to 375°F. Line 2 large baking sheets with foil or parchment paper.

- With a sharp knife, cut the log into ¼"-thick slices. Place on the prepared baking sheets, leaving 1" space between cookies. Bake for 7 minutes, or until lightly browned. Let cool on a wire rack. Continue until all cookies are baked.

KITCHEN TIP

To freeze, place the cooled cookies on baking sheets and freeze for 30 minutes, or until solid. Transfer to freezer-quality plastic bags. To thaw the whole batch to have on hand, place overnight in the refrigerator. To thaw cookies to eat right away, place the number you need on a baking sheet and bake at 350°F for 10 minutes, or until warm. For crisper cookies, bake for 12 to 15 minutes.

THE FINISHING TOUCH

Cakes, cupcakes, cookies, and other baked goods freeze best without glazes or frosting, since these toppings tend to separate and run after thawing. You can, however, whip up very quick glazes and frostings to crown your sweet freezer treats. Each of these recipes covers one dozen cookies or cupcakes or an 8" × 8" cake and costs 79¢ less than a comparable amount of commercial frosting.

- *Chocolate Decadence.* In a small bowl, combine 1¾ cups confectioners' sugar, 3 tablespoons skim milk, 2 tablespoons cocoa powder, and ½ teaspoon vanilla. Mix well.

- *Light and Lemony.* In a small bowl, combine 1¾ cups confectioners' sugar, 2 tablespoons skim milk, 1 teaspoon lemon juice, and ½ teaspoon grated lemon rind. Mix well.

- *Orange Glaze.* In a small bowl, combine 1¾ cups confectioners' sugar, 2 tablespoons skim milk, and 1 drop orange extract or 1 tablespoon grated orange rind. Mix well.

Per serving
Calories 267
Total fat 3.6 g.
Saturated fat 1.8 g.
Cholesterol 7 mg.
Sodium 45 mg.
Fiber 5.5 g.

Cost per serving

64¢

RASPBERRY CRISP

This unusual combination of raspberries and peaches makes a vivid and delicious crisp for only 64¢ a serving. Top with low-fat vanilla frozen yogurt.

1   orange
3   cups frozen sliced peaches
3   cups frozen raspberries
½   cup packed light brown sugar
1   cup all-purpose flour
½   cup rolled oats
½   teaspoon ground cinnamon
2   tablespoons reduced-calorie butter
⅓   cup low-fat buttermilk

❋ Preheat the oven to 350°F. Coat an 8″ × 8″ baking dish with no-stick spray.

❋ Grate the rind from the orange into a medium bowl. Squeeze the orange juice into the bowl. Add the peaches, raspberries, and ¼ cup of the brown sugar. Mix well. Spoon into the prepared baking dish.

❋ In a 10″ no-stick skillet, combine the flour, oats, and cinnamon. Cook, stirring, over medium heat for 3 to 5 minutes, or until light brown. Remove from the heat. Stir in the butter, buttermilk, and the remaining ¼ cup brown sugar. Spoon over the fruit.

❋ Bake for 45 minutes, or until the topping is golden and the fruit is bubbling.

Fruit Compote with Sugared Wonton Cookies

Makes 4 servings

Per serving
Calories 227
Total fat 1.3 g.
Saturated fat 0.3 g.
Cholesterol 1 mg.
Sodium 53 mg.
Fiber 8.4 g.

Cost per serving

$1.39

Wontons are a convenient item to have in the freezer. You can freeze the wrappers in the original package. If you need only a few wontons for a recipe, thaw the package for about 30 minutes at room temperature. Peel off the number of wontons you need and then rewrap the remaining number tightly in freezer-quality plastic wrap or foil. Wontons can be refrozen 2 or 3 times without damaging their texture.

2 cups frozen blackberries
2 cups frozen blueberries
1 cup sliced bananas
1 cup sliced frozen peaches
2 tablespoons maple syrup
1 tablespoon chopped fresh mint
1 teaspoon lime juice or lemon juice
4 frozen wonton wrappers, thawed
2 tablespoons sugar

✳ Preheat the oven to 425°F.

✳ In a large bowl, combine the blackberries, blueberries, bananas, peaches, maple syrup, mint, and lime juice or lemon juice. Let stand at room temperature, stirring occasionally, for 20 minutes.

✳ Coat both sides of the wontons with no-stick spray. Sprinkle with the sugar. Place on a baking sheet. Bake for 5 minutes. Turn and bake for 2 minutes, or until golden brown and crisp. Place the wontons on 4 dessert plates. Top with the fruit mixture.

Kitchen Tip

Get full value from any bunch of herbs you purchase. After using the amount you need for a recipe, chop and freeze the remaining leaves in a small freezer-quality plastic bag. Mint and other herbs will remain green and flavorful. Use the herbs while still frozen.

FRUITFUL GAINS

When you score big at your local farmers' market, consult these simple guidelines to prepare fruit for the freezer. Pack the fruit in freezer-quality plastic bags or containers.

❀ *Apples*. Core, peel, and slice. Toss with a small amount of honey or lemon juice to prevent browning. (1¼ pounds whole apples equals about 1 pint slices)

❀ *Apricots*. Peel ripe apricots. Dip slightly underripe apricots in boiling water for 30 seconds, then plunge into cold water; remove the peels, then halve and pit. Toss with a small amount of honey mixed with an equal amount of water to prevent browning. (1⅔ pounds whole apricots equals about 1 pint halves)

❀ *Cherries*. Pit before freezing. (1½ pounds equals 1 pint)

❀ *Grapes*. Remove from stems and pack whole. (1½ pounds equals 1 pint)

❀ *Peaches*. Peel ripe peaches. Dip slightly underripe peaches in boiling water for 30 seconds, then plunge into cold water; remove the peels, then halve and pit. Toss with a small amount of honey and lemon juice to prevent browning. (1⅔ pounds whole peaches equals about 1 pint halves)

❀ *Pears*. Peel, core, and slice. Toss with a small amount of lemon juice to prevent browning. (1 pound whole pears equals about 1 pint slices)

❀ *Plums*. Halve, pit, and puree. (1 pound whole plums equals 1 pint puree)

APPLE BROWN BETTY

You can serve this dessert everyday-style by spooning it into bowls or upgrade it to fancy-style by inverting the entire dessert onto a pretty serving plate.

4 tart apples, peeled and sliced
2 tablespoons lemon juice
2 tablespoons honey
2 cups soft bread crumbs
¼ cup reduced-calorie butter, at
    room temperature
¼ cup packed brown sugar
½ teaspoon ground cinnamon
¼ teaspoon ground allspice

❉ Preheat the oven to 375°F. Coat a 9″ × 5″ loaf pan with no-stick spray. Lightly flour the pan and shake off any excess.

❉ In a medium bowl, combine the apples, lemon juice, and honey. Toss to coat. Transfer to the prepared pan.

❉ In the same bowl, combine the bread crumbs, butter, brown sugar, cinnamon, and allspice. Mix well. Arrange on top of the apples. Bake for 40 minutes, or until the apples are soft and the topping is crisp.

ROLLING IN DOUGH

Commercially refrigerated cookie dough can cost as much as 15¢ to 25¢ a cookie, while a homemade cookie costs only 3¢. So it pays to spend 30 minutes on a weekend to make multiple batches of your family's favorite cookie dough to have on hand for desserts, lunch boxes, company, or gifts. Just visualize the coins stacking up with each cookie that's baked.

To freeze cookie dough, form it into logs about 2 inches in diameter. Wrap with freezer-quality plastic wrap, then pack into labeled freezer-quality plastic bags. Freeze for up to 3 months. Thaw each log overnight in the refrigerator or at room temperature for 1 hour, then slice and bake according to the recipe.

Per serving
Calories 271
Total fat 8.6 g.
Saturated fat 5.2 g.
Cholesterol 20 mg.
Sodium 189 mg.
Fiber 3.2 g.

Cost per serving

46¢

KITCHEN TIP

To freeze, invert the brown betty onto a freezer-proof plate and let cool. Wrap with freezer-quality plastic wrap, then freezer-quality foil. To use, thaw overnight in the refrigerator. Reheat in a 375°F oven for 15 minutes, or until hot.

Per serving
Calories 190
Total fat 0.6 g.
Saturated fat 0.1 g.
Cholesterol 0 mg.
Sodium 40 mg.
Fiber 3.2 g.

Cost per serving

77¢

KITCHEN TIP

To freeze, place the cooled meringue shells on a baking sheet in the freezer for 30 minutes, or until the meringues are solid. Transfer the shells to a tin or freezer-quality plastic container. To use, thaw at room temperature for 30 minutes.

MERINGUE SHELLS WITH MIXED BERRIES

With the aid of the freezer, serving this elegant dessert to company is nearly effortless. You can prepare the crisp yet cloudlike meringue shells and freeze them for up to 3 months. Then you can pull together the berry filling just before serving.

Meringue Shells

4 egg whites, at room temperature
⅛ teaspoon cream of tartar
½ teaspoon vanilla
¼ cup sugar

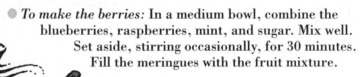

FROM the Freezer

Berries

2 cups frozen blueberries
2 cups frozen raspberries
1 tablespoon chopped fresh mint
¾ cup sugar

❋ *To make the meringue shells:* Preheat the oven to 225°F. Line a large baking sheet with foil or parchment paper.

❋ Place the egg whites in a large bowl. Beat with an electric mixer until foamy. Add the cream of tartar and beat until soft peaks form. Beat in the vanilla, then gradually beat in the sugar and continue beating until stiff peaks form.

❋ Mound in 4″ circles on the prepared baking sheet. With the back of a spoon, make an indentation in the center of each circle. Bake for 1½ hours, or until the meringues are dry and crisp. Let cool on the baking sheet.

❋ *To make the berries:* In a medium bowl, combine the blueberries, raspberries, mint, and sugar. Mix well. Set aside, stirring occasionally, for 30 minutes. Fill the meringues with the fruit mixture.

Honeyed Tapioca Pudding with Peaches

It's comforting to have the old-fashioned goodness of tapioca pudding tucked into the freezer for a fast dessert.

⅓ cup packed brown sugar
¼ cup quick-cooking tapioca
¼ cup honey
2 eggs
¼ teaspoon ground nutmeg
2 cups skim milk
3 cups frozen sliced peaches, thawed
1 teaspoon lemon juice
2 tablespoons chopped crystallized ginger

❋ In a medium saucepan, combine the brown sugar, tapioca, honey, eggs, and nutmeg. Whisk until smooth. Whisk in the milk. Cook over medium heat, whisking constantly, for 5 minutes, or until the mixture comes to a boil. Remove from the heat.

❋ Partially fill a large bowl with ice. Set the saucepan on the ice and let stand, stirring occasionally, for 15 minutes, or until the tapioca is cool.

❋ In a medium bowl, toss the peaches with the lemon juice. Arrange in 4 serving dishes. Spoon the tapioca over the peaches. Top with the ginger.

Kitchen Tip

To freeze, pack the cooled tapioca into a freezer-quality plastic container. To use, thaw overnight in the refrigerator. Prepare the peaches and ginger just before serving.

BERRY MERRY

You can pick incredible savings along with fresh berries from June through September. Fresh blueberries drop as low as 50¢ per pint, about one-third the price you'd pay for a comparable amount of frozen blueberries in the supermarket.

Here's how to prepare berries for freezing. Pack them in freezer-quality plastic bags or freezer-quality plastic containers.

Blackberries. Freeze whole on a baking sheet before packing. Or mash in a bowl with a small amount of sugar. Freeze for up to 4 months.

Blueberries. Freeze whole on a baking sheet before packing. Freeze for up to 4 months.

Currants. Freeze whole on a baking sheet before packing. Freeze for up to 12 months.

Raspberries. Freeze whole on a baking sheet before packing. Freeze for up to 4 months.

Strawberries. Freeze whole on a baking sheet before packing. Or slice, then mash with a small amount of sugar. Freeze for up to 4 months.

BLUEBERRY WHIPPED DESSERT

Freezing ricotta is an ideal way to save small amounts left over from making a recipe. Just remember to thaw the ricotta overnight in the refrigerator the night before making this dessert.

Makes 4 servings

Per serving
Calories 132
Total fat 3 g.
Saturated fat 1.6 g.
Cholesterol 10 mg.
Sodium 64 mg.
Fiber 2.1 g.

Cost per serving

67¢

2 cups frozen blueberries
1 tablespoon packed brown sugar
1 teaspoon cornstarch
1 teaspoon lemon juice
½ cup part-skim ricotta cheese
½ cup nonfat plain yogurt
2 tablespoons confectioners' sugar
1 teaspoon vanilla

❋ In a small saucepan, combine the blueberries, brown sugar, cornstarch, and lemon juice. Cook, stirring, over medium-high heat for 5 minutes, or until the sauce thickens slightly. Transfer to a large bowl, cover, and refrigerate until chilled.

❋ In a blender or food processor, combine the ricotta, yogurt, confectioners' sugar, and vanilla. Process until smooth. Fold into the blueberry mixture. To serve, spoon into 4 dessert bowls.

KIWI-STRAWBERRY PARFAITS

The ingredients for this easy dessert are both fresh and frozen. You assemble it in minutes, then pop it back into the freezer until serving time.

Makes 4 servings

Per serving
Calories 280
Total fat 1.3 g.
Saturated fat 0.4 g.
Cholesterol 3 mg.
Sodium 38 mg.
Fiber 5.6 g.

Cost per serving

$1.54

6 cups frozen sliced strawberries
½ cup frozen raspberries
1 cup frozen low-fat vanilla frozen yogurt
½ cup sugar
4 kiwifruits, sliced

❋ In a blender or food processor, combine the strawberries, raspberries, yogurt, and sugar. Process until smooth.

❋ Divide half of the kiwifruit among 4 parfait glasses. Fill each glass halfway with the strawberry mixture. Repeat with the remaining kiwifruit and strawberry mixture. Freeze for 30 minutes, or until the mixture is slushy.

Per serving
Calories 285
Total fat 1.9 g.
Saturated fat 1 g.
Cholesterol 1 mg.
Sodium 19 mg.
Fiber 5.4 g.

Cost per serving
37¢

BANANA ICE CREAM WITH CHOCOLATE SAUCE

Overripe bananas are a great freezer resource. Peel before freezing and use them for this quick and luscious dessert that's reminiscent of a hot-fudge banana split.

¾ cup sugar
7 tablespoons cocoa powder
1 teaspoon cornstarch
½ cup skim milk
4 frozen overripe bananas

✻ In a medium saucepan, whisk together the sugar, cocoa, cornstarch, and ¼ cup of the milk to form a paste. Place over medium heat. Stir in the remaining ¼ cup milk. Cook, stirring, over medium heat for 5 minutes, or until smooth and slightly thickened. Let cool for 5 minutes.

✻ Cut the bananas into 1″ pieces. Place in a blender or food processor. Process until smooth.

✻ Spoon the banana ice cream into serving dishes. Top with the chocolate sauce.

Per serving
Calories 119
Total fat 0.2 g.
Saturated fat 0 g.
Cholesterol 0 mg.
Sodium 10 mg.
Fiber 2.7 g.

Cost per serving
55¢

STRAWBERRY ICE MILK

This pretty dessert takes only 5 minutes of hands-on time.

2 cups frozen strawberries
2 cups frozen sliced peaches
4 ice cubes
¼ cup sugar
¼ cup skim milk
¼ teaspoon almond extract or 1 teaspoon vanilla

✻ In a blender or food processor, combine the strawberries, peaches, ice cubes, sugar, milk, and ½ teaspoon of the almond extract or vanilla. Process until smooth. Taste and add up to ½ teaspoon more almond extract or vanilla if needed. Process to combine. Serve immediately.

RECIPES FOR/FROM THE FREEZER

Recipes
For/From
the Freezer

INDEX

Note: <u>Underscored</u> page references indicate boxed text.

P

International Conversion Chart

These equivalents have been slightly rounded to make measuring easier.

Volume Measurements

U.S.	Imperial	Metric
¼ tsp.	–	1.25 ml.
½ tsp.	–	2.5 ml.
1 tsp.	–	5 ml.
1 Tbsp.	–	15 ml.
2 Tbsp. (1 oz.)	1 fl. oz.	30 ml.
¼ cup (2 oz.)	2 fl. oz.	60 ml.
⅓ cup (3 oz.)	3 fl. oz.	80 ml.
½ cup (4 oz.)	4 fl. oz.	120 ml.
⅔ cup (5 oz.)	5 fl. oz.	160 ml.
¾ cup (6 oz.)	6 fl. oz.	180 ml.
1 cup (8 oz.)	8 fl. oz.	240 ml.

Weight Measurements

U.S.	Metric
1 oz.	30 g.
2 oz.	60 g.
4 oz. (¼ lb.)	115 g.
5 oz. (⅓ lb.)	145 g.
6 oz.	170 g.
7 oz.	200 g.
8 oz. (½ lb.)	230 g.
10 oz.	285 g.
12 oz. (¾ lb.)	340 g.
14 oz.	400 g.
16 oz. (1 lb.)	455 g.
2.2 lb.	1 kg.

Length Measurements

U.S.	Metric
¼"	0.6 cm.
½"	1.25 cm.
1"	2.5 cm.
2"	5 cm.
4"	11 cm.
6"	15 cm.
8"	20 cm.
10"	25 cm.
12" (1')	30 cm.

Pan Sizes

U.S.	Metric
8" cake pan	20 x 4-cm. sandwich or cake tin
9" cake pan	23 x 3.5-cm. sandwich or cake tin
11" x 7" baking pan	28 x 18-cm. baking pan
13" x 9" baking pan	32.5 x 23-cm. baking pan
2-qt. rectangular baking dish	30 x 19-cm. baking pan
15" x 10" baking pan	38 x 25.5-cm. baking pan (Swiss roll tin)
9" pie plate	22 x 4 or 23 x 4-cm. pie plate
7" or 8" springform pan	18 or 20-cm. springform or loose-bottom cake tin
9" x 5" loaf pan	23 x 13-cm. or 2-lb. narrow loaf pan or paté tin
1½-qt. casserole	1.5-liter casserole
2-qt. casserole	2-liter casserole

Temperatures

Farenheit	Centigrade	Gas
140°	60°	–
160°	70°	–
180°	80°	–
225°	110°	–
250°	120°	½
300°	150°	2
325°	160°	3
350°	180°	4
375°	190°	5
400°	200°	6
450°	230°	8
500°	260°	–